SCIENCE EXPERIENCES FOR THE EARLY CHILDHOOD YEARS

FIFTH EDITION

JEAN D. HARLAN

Merrill, an imprint of
Macmillan Publishing Company
New York

Maxwell Macmillan Canada
Toronto

Maxwell Macmillan International
New York Oxford Singapore Sydney

Cover art/photo: J. nes Levin/FPG International
Editor: Linda A. Su. 'van
Production Editor: Sh. ryl Glicker Langner
Art Coordinator: Lorra. e Woost
Cover Designer: Robert ga
Production Buyer: Pamela . Bennett

This book was set in Palatin. by Carlisle Communications, Ltd. and was printed and bound by R. R. Donnelley & Sons Company. The cover was printed by New England Book Components.

Printed in the United States of America

Macmillan Publishing Company
866 Third Avenue
New York, NY 10022

Macmillan Publishing Company is part of the
Maxwell Communication Group of Companies.

Maxwell Macmillan Canada, Inc.
1200 Eglinton Avenue East, Suite 200
Don Mills, Ontario M3C 3N1

Library of Congress Cataloging-in-Publication Data
Harlan, Jean Durgin.
Science experiences for the early childhood years / Jean D. Harlan.—5th ed.
p. cm.
Includes bibliographical references and index.
ISBN 0-02-350170-7
1. Science—Study and teaching (Elementary) I. Title.
LB1585.H297 1992
372.3'5044—dc20 91-19841
CIP

Printing: 1 2 3 4 5 6 7 8 9 Year: 2 3 4 5

Preface

The spotlight of White House concern continues to focus on the disheartening national crisis in science education. The crisis emcompasses declining science achievement test scores by today's students, and as predicted, future shortages of scientists and science educators in a period of serious global environmental problems. Recent national surveys of the general level of science literacy reveal another result of long-term weak science education: a large segment of our population is scientifically illiterate. Environmental renewal and protection efforts in recent decades have fallen short, in part because this ill-informed constituency was not aroused by the urgency of these problems. The global concerns were not understood in terms of basic laws of nature.

As one remedial step, the U.S. Department of Education has set a goal of raising our high school students' science achievement levels to rank with the highest international scores by the year 2000. Readers of this textbook currently are, or soon will be, teaching the young children who are targeted to be those turn-of-the-century high school students. Each of you has a potential role in resolving the crisis in science education. The revisions in this textbook were planned to help you with your foundation-building task.

This fifth edition reports current cognitive research on the social aspects of learning. These findings support the original premise that active learning is enhanced by appropriate guidance from teachers.

In keeping with the cognitive/affective rationale for this book, new teaching resources and references have been included to help anxious youngsters cope with powerful and frightening natural events. Information about earthquakes, floods, and violent storms provides children with the only control they can have over their fears: the reassurance of planning for safety if such a disaster occurs.

This edition clarifies criteria for selecting the topics for early childhood science. New references are listed for two popular topics that some teachers may be able to include in their plans. Two new imagery experiences have been added to the holistic extensions in the activities section of this book. Also, stronger emphasis has been placed on ecological relationships and environmental concerns. Finally, in consonance with the federal Project Education goals, added attention has been given to the important part parents and families play in strengthening children's interest in science. Enjoyable *Exploring at Home* suggestions for home-based science activities have been added for each science topic. Resources for parents are listed.

To revitalize science education, both teaching and learning must be invested with all of the fascination and enthusiasm it can generate in fully involved participants. The intent of this book is to inspire readers to help children learn what they ardently wish to know: how the physical and natural worlds really work.

ACKNOWLEDGMENTS

It is my good fortune to have many contacts, both informally and professionally, with fine classroom teachers and with eager young learners. These associations sustain my efforts to keep this textbook current. My revision was made easier because of helpful conversations and correspondence with Dr. Mary Rivkin, University of Maryland, Baltimore County, and with my co-consultant in science education for *Sesame Street,* Dr. Steven Tipps, Midwestern State University. I appreciate their insights and their encouragement. Helpful comments were also offered by the following reviewers: Audrey Boyd, Guilford Technical Community College; Adolfo Sanchez, Cayey University College (Puerto Rico); and Charles Mitchell, SUNY Plattsburgh.

Susan M. Sorenson added a new illustration to those of my good friend, Anne Clark Culbert, a fine naturalist, writer, and artist. Finally, for their positive influence on my life, I am deeply grateful to my aunts, Faith Palmer and Fran Trumbull; to my children: Betsy Bales, Anne Strohm, John Harlan, M.D., Susan Borghese, and Julie Harlan; and to my delightful grandchildren: Katie, Rachel, Liz, Christopher, Lauren, and Nina. Thank you all.

Contents

PART ONE

THE RATIONALE

1

A Cognitive/Affective View of Science Learning

AFFECTIVE COMPONENTS OF LEARNING

"Wow! I just discovered electricity!"

That was Russell, leaping from his chair to shout the results of his patient struggle with battery, wires, and bulb. Though many others preceded him in harnessing electricity, Russell truly discovered for himself the path to a glowing bulb and personal exultation.

How can a 5-year-old like Russell concentrate for 15 minutes on the challenge of completing a simple circuit? Why, for that matter, do children want to know why? One part of the answer is that understanding the environment by interacting with it is the natural work of young children. From birth onward, infants use and slowly coordinate their reflexes and perceptions into information-gathering sources. The early looking, grasping, and mouthing of objects gradually build into specific ways to recognize and handle different materials. Simple discoveries of cause and effect allow babies to solve playpen problems. If curiosity and exploring are encouraged, the probing for meaning broadens as language and locomotion develop. By giving children opportunities and support, their inherent need to know continues to deepen and serves as part of the motivation for more complex investigations such as Russell's (White, 1968).

When the human desire for understanding the world is organized into careful ways of collecting, testing, and sharing information, it is called *science*. Indeed, the word *science* is derived from the Latin root *scire*, meaning "to know." When we offer intriguing science experiences to young children, we nourish their natural, human capacity to know. If we do this with sensitivity to their interests, nature, and needs, we release and enhance the powerful affective component of knowing and learning.

The affective component of knowing and learning is a complex web of interrelated facets, including curiosity and emotional responses to life experiences, and the self-esteem that stems from becoming competent. The way children feel about themselves and their world influences their curiosity. There is an energizing, reciprocal link between finding out and self-esteem, between feelings of mastery of the newly learned and the desire to know more. Anxious concerns about unexplained or frightening events may lure children into finding answers. Pleasure in discovering awesome, delightful, or comforting features of the environment also furthers affective and cognitive growth. This chapter will consider some effects of these interrelationships, their biological basis, and multimodal ways to use these interrelationships and biological givens to enhance science learning.

Curiosity and Emotions

Many of the existing theories about curiosity view it not as an emotion in itself, but as an *affect:* a mental state that influences emotions. Researchers have also demonstrated that emotions, in turn, have a strong influence on curiosity. In classic studies, Harlow (1958) analyzed the spontaneous curiosity and exploratory behavior of infant monkeys, and Ainsworth and Bell (1970) observed that of human infants. Both investigators confirmed that feelings of security enable curiosity and exploration, while insecurity and strong fear can interrupt and even paralyze curiosity. Other studies by ethologists add that, while overwhelming fear prevents exploring, uncertainty tinged with a bit of fear of the unknown seems to stimulate curiosity and exploring.

Emotions and Learning Interest

It is easy to observe the lasting interest in learning that can result from a warm firsthand experience. This was the case when a group of children in a day care center unexpectedly discovered an untended nest of young rabbits in their play yard. The children were involved in the efforts to keep the babies fed and sheltered until they were old enough to leave the nest. For many months, the children spoke about "their" rabbits. They responded eagerly to new information about rearing animal young. The pleasure and pride they gained from nurturing the rabbits provided an emotional link to the new learning. They demonstrated one way that mutually enhancing cognitive/affective processes lead to meaningful learning.

Occasionally the connection between a child's anxious concern and his or her eagerness to find answers is clearly apparent. In this example, the mother of 4-year-old Matthew asked for advice on how to help her son get to sleep at night. He was unable to shut his eyes until he had counted all the stars he could see from his window. He needed to count them over and over to make sure they were still all in place. That provided reassurance for him that a star wouldn't fall on him while he was sleeping.

In response to his concern, a series of experiences was provided to build understanding about why things fall to earth, or in this case, why they don't fall. Matthew enjoyed some tangible explorations of the effects of magnetism, an invisible force that could, nonetheless, be felt in the hands of a fascinated boy. After many experiences with magnets, Matthew experimented with some simple effects of the greater invisible force of gravity. He found that the hardest toss by a strong boy couldn't send a ball beyond the pull of the earth's gravity. Even he himself always came back to earth after his highest leaps. Matthew helped build a carton spaceship to play at being an astronaut, floating so far from earth that earth's gravity could not pull him down. He frequently asked to be read a story about astronauts who traveled 3 days to reach the moon, because it was so far away. He always chimed in at the part telling that stars were even farther away from earth than the moon.

The day finally came when Matthew's mother reported that it was now easier for him to get to sleep. He only had to do one quick scan of the stars before climbing into bed. He explained confidently to his parents that earth's gravity isn't strong enough to pull down a star. In the years since then, Matthew's fascination with astronomy has continued. He and his family have made several visits to an observatory and a planetarium in a distant city, and he now owns a small telescope. The story of Tony in Selma Fraiberg's *The Magic Years* (1950) is a similar example of how a child's fear led to an enduring interest in science.

Problem-Solving, Competence, and Self-Esteem

Science experiences have special potential for building a sturdy concept of the self as a person who can cope with problems. Martin Seligman's studies (1975) have led him to conclude that the belief in one's competence begins in infancy and develops throughout life as mastery motivation. He believes that if a young child is not provided with, or is not allowed to cope with, problems that can be resolved through his own actions, a pattern of helplessness begins. Seligman feels that too-easy success, as well as too-little challenge, produces children with a limited capacity to cope with failure. Therefore, he calls for offering learning challenges in school against which children can measure themselves, since challenge is a source of a person's sense of self-worth.

According to Seligman, our self-esteem and sense of competence do not depend so much on whether good or bad things happen to us, but on whether we believe we have some control over what happens to us. Early science experiences can provide children with some sense of control through the ability to predict certain things that will happen; for example, "The pan of snow will change to water if we keep it indoors." In this way, some of the confusion and uncertainty in events occurring around the child can be replaced with that awareness of predictability. The child can learn that even she herself can bring about some of the small occurrences. Science knowledge helps develop the only control a child can have over powerful and sometimes unpredictable natural forces: control of attitude through understanding the causes of the event and control of helpless fear through training to cope safely with events such as earthquakes and violent storms. Science experiences are unique in fostering this strengthening influence on the child's personality.

BIOLOGICAL BASIS OF THE COGNITIVE/ AFFECTIVE LINK

The intimate link between feelings and thinking has long been discerned. Jean Piaget is among those who have offered insights about this interrelatedness. He pointed out that the affective, social, and cognitive aspects of behavior are inseparable (Kamii & DeVries, 1978). In her influential work on nature study, Rachel Carson wrote, "Once the emotions have been aroused, then we wish for knowledge about the object of our emotional response. Once found, it has lasting meaning" (1960, p. 45). The reverse was also noted by Richard Restak (1979), who said, "If emotions (associated with an event) are *muted,* memories of the event are less enduring" (p. 194).

Contemporary anatomical and physiological descriptions of certain brain functions show us that there is a biological basis for the interrelated effects we have observed. One way of understanding this link is to view the brain as having three separate structures, each differing in nature and function:

1. ***The Lower Brain:*** Located at the inner base of the brain, this structure directs, via the central and peripheral nervous systems, the automatic life processes that keep the body alert and the heart and lungs functioning.
2. ***The Limbic System:*** (mid-brain) This structure overlies the lower brain. One part of the limbic system, the hippocampus, plays an important part in the ability to recall. Incoming data from the senses pass through the limbic system close to the hippocampus. The hypothalamus, also part of the limbic system, secretes chemical messengers (hormones and neurotransmitters) in response to emotions. Depending on their type or quantity, the neurotransmitters can either intensify or suppress memory. Mild stress and anxiety release the hormone adrenaline, which facilitates recall. However, high levels of anxiety produce *excessive* outpourings of adrenaline that *reduce* recall (Restak, 1979). Pleasurable feelings and perception of novelty and complexity stimulate this area of the brain (Sagan, 1977).
3. ***The Cerebrum:*** This most complex of the structures makes up about 85 percent of the total brain area. The higher learning centers are located in the outer surface of this structure, the neocortex (cerebral cortex). Here incoming data are processed and stored, and associations are made. Different areas of the neocortex serve highly specialized functions such as language understanding and production. The neocortex can be considered the "executive" part of the brain, but the information it receives and the responses it sends must pass through the limbic system and the lower brain.

The functions of the three structures of the brain show how the cognitive, affective, and psychomotor responses are interconnected. The limbic system influences the functioning of the other two brain structures through chemical secretions. Millions of nerve connections between the limbic system and the cerebral cortex and within the cerebral cortex are the sites of cognitive processing. The chemical and neural communication systems make the cognitive/affective interrelationships possible. With this information we can now conclude that the intermeshing of feelings and thinking is not a random happening: rather, it is a biological phenomenon.

MULTIMODAL LEARNING ENHANCEMENT

Children confirm and expand science concepts through a variety of integrating activities. The practice of extending learning through different modes has been followed by nursery and progressive school teachers for half a century. It has been shown in past research that learning is facilitated when ideas are presented to young children in a variety of contexts (Razran, 1961). Now, the Nobel Prize-winning research of Roger Sperry (1970) has provided physiological evidence that explains why this is so. His work identified distinct functions of the two hemispheres of the brain and the multiple neural pathways between them.

Brain Hemisphere Functioning

The cortex is divided into right and left hemispheres, each of which processes information in different ways. For many people, the left brain appears to specialize in step-by-step, sequential, symbolic, often language-related thought. The right hemisphere seems to work in a more patterned, simultaneous relationship fashion, using images, analogies, metaphors, and spatial connections. The right hemisphere tends to combine images, experiences, and emotions in a way that encourages invention. The rapidity of these resulting ideas and perceptions makes them seem as though they have not been reasoned out; hence they are considered to be intuitive.

According to Steven Tipps (1990), the differences in processing styles of the two hemispheres is often overgeneralized into dichotomies, such as logical versus creative and linguistic versus mathematical. He says the most important idea about hemisphericity is that people have two ways of learning, knowing, and remembering.

The two sides of the brain do not operate in isolation, but in conjunction with each other (Languis, Sanders, & Tipps, 1980). While each half of the brain handles different aspects of meaning from the same experience, multiple neural connections between the brain halves make it possible for the two processing styles to evaluate and integrate information simultaneously. Putting creative thinking to use is an example of integrating right- and left-brain functioning. Both the ability to see a new relationship and the ability to plan what to do with that relationship are necessary to solve problems, whether scientific or everyday problems.

The potential of the brain is best used when both the right- and left-brain processes are engaged. Traditional teaching techniques of lecture and recitation may appeal only to the linear, logical approach of the left hemisphere and limit deeper learning. Learning experiences that activate both hemispheres double the potential for learning and creating.

Two Ways of Knowing

Carl Sagan (1977) points out that the distinctly different types of information processing occurring in the two brain hemispheres explain and validate the two different ways we have of knowing:

1. *Rational knowledge* is verbal and systematic, originating in the left hemisphere. We typically consider scientific thinking as primarily rational thinking.
2. *Intuitive knowledge* comes from rapid, nonverbal perceptions that occur primarily in the right hemisphere by synthesizing feelings and perceptions from both sides of the brain. Enlightening, creative insights seem to come to us unbidden, often during periods of stress-free relaxation.

Both forms of knowledge can be accurate, and each form can enhance the other. Educators in Western cultures have tended to respect only rational thought that can be promoted through formal teaching. Intuitive thought has been distrusted and dismissed as being less verifiable. It seems to be considered less important because it comes to mind too quickly to be explained. Piaget recognized intuitive thinking in children's mental life; but he, too, felt that the intuitive knowing was not valid understanding because it could not be explained verbally by logical reasoning.

However, intuitive (creative) insights can be essential to scientific thinking, both in identifying problems to be solved and in making hunches about how to logically pursue a solution. For centuries there has been evidence of the role intuition and imagery have played to inspire great scientific discoveries, as well as master works in the arts and humanities (Harman & Rheingold, 1984).

Groundbreaking solutions to major problems like the production of insulin and the first model of atomic structure have come intuitively to scientists during periods of relaxation and sleep. This process also occurs in our everyday thinking as we struggle to understand how to use a new piece of equipment or try to resolve a stubborn problem. The solution often begins intuitively with "Oh, now I see. . . ." and a useful fresh approach opens to us.

The notion of incorporating creative thinking and imagery into science education is no longer being challenged as it was even a decade ago. It is acknowledged that both intuitive and rational thinking function together for optimal use of intelligence to generate creative ideas. Now there is a growing interest in finding ways to promote intuitive/creative thinking in the classroom (Clark, 1986).

We encourage intuitive/creative thinking when we respect spontaneous demonstrations of children's intuitive and rational thought working together. Andrew offered a charming example during a class discussion of the story, "Little Bear Goes to the Moon," by Minarek. We were speculating about what would happen when Little Bear, pretending to fly to the moon, actually jumped from a tree stump. Andrew's prompt response was, "See, air and gravity are fighting. Air wants to hold Little Bear up, so he will float. But he's too heavy, so gravity will win and pull him down." Andrew recalled a concept about air he

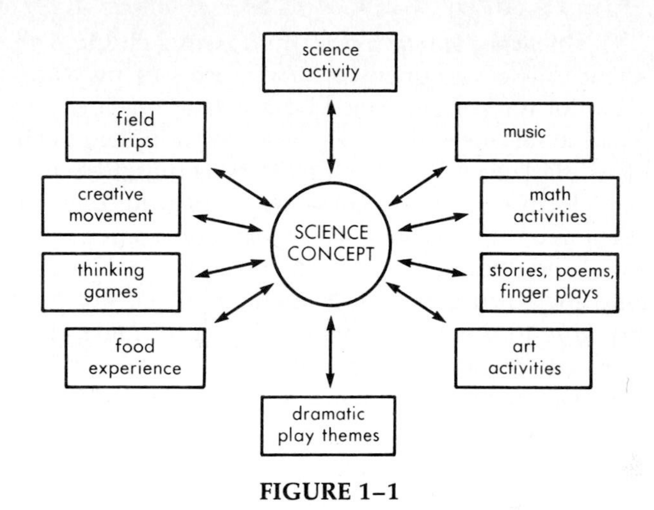

FIGURE 1–1

had acquired five months earlier. He integrated this idea with his hunch about gravity's effect on Little Bear, and expressed his ideas in a metaphor familiar to this aggressive 5-year-old.

MULTIMODAL LEARNING

Primary Integrating Activities

Children acquire science concepts through direct observation and exploration of familiar events. The concepts are best understood when they are presented in a variety of ways that relate to other events in their world and to other learnings in the classroom. Core science concepts are more durably retained when they are reinforced through creative and psychomotor ways of learning. A variety of integrating activities are suggested for each of the major science topics presented in Part Two of this book. The effect is diagrammed in Figure 1–1.

Art activities are suggested to engage the spatial and intuitive style of thinking as well as the planning and logic to complete the expression of ideas. Drawing, painting, and modeling call for spatial thinking and focused concentration as children create personal interpretations of science events. Art projects using materials that were part of the science activities encourage divergent thinking as children invent new ways to use the materials.

Poems, fingerplays, and stories extend science experiences by associating them with imagery and language. Poetry evokes images, while stories also provide additional confirmation of the regularity of physical or natural events. Whether embedded in fiction, or appearing as central themes of science-based stories, the ideas register reassuringly in children's minds as familiar knowledge. Creative thinking is promoted when children are encouraged to produce their own science-theme stories and poems.

Dramatic play situations afford young children chances to test out and apply science ideas imaginatively. Two forms of play are suggested: creative drama ideas for impromptu guided dramatization of known stories; and spontaneous play themes for which a few props are supplied to stimulate children's own play ideas. Barbara Biber (1979) reminds us that dramatic play has been considered a valid form of learning since the pioneering days of nursery school, nearly 70 years ago: "It was seen as a means of deepening insights, integrating knowledge, and finding identification on a personal level." Recent research shows that play and science are complementary aspects of problem solving. Science gives structure to the activity, while play encourages creative behaviors and positive attitudes toward solving problems (Severide & Pizziui, 1984). Note, however, that any play loses its value as pleasurable activity if adults hover closely, ever ready to step in to capitalize on "teachable moments."

Creative movement uses the spatial form of knowing, as well as the motor pathways to learning. Physical encoding (forming mental symbols) takes place as abstract ideas are intuitively translated into concrete physical movements of the body. Both understanding and retention of ideas are increased in this joyful, relaxing way.

Music can provide cues to recall information, since hearing, like all of our sensory systems, evokes memory. As tunes run through the children's minds, they are also reminded of the ideas expressed in the lyrics. Catchy words, rhythms, and sounds are standard means of luring customers to buy products because of their power as memory cues.

Thinking games make it fun for children to recall science information. Steven Tipps (1982) values strategy games that do more than allow for memorization of facts. "As children play games many times, they confront the information in different orders and different contexts." He points out that simple card games such as Bingo, Go Fish, and concentration can be adapted to almost any content. Other mind stretchers encourage intuitive thought by involving the imagination in "what if . . . " fantasy.

Math activities are an integral part of all science as means of quantifying and recording observations. Quantifying is a means of knowing the world that seems to occur naturally to young humans. Some of the math activities suggested in later chapters are necessary to the science activities as a way to gather data. Others use science themes to provide a new context for using math skills.

Food experiences use taste and smell to strengthen the recall of concepts. The pleasure of being involved in preparing or sampling good things to eat provides a heightened emotional state for lasting memories. For all of us there are vivid connections between particular foods and our memory of the feelings we associated with those foods. Edible science experiences can be deliberately used to strengthen concept retention.

Field trips add importance to ideas acquired in science activities about the behavior of matter and living things. Concepts are validated and understood in a new way when they can be seen in action. Children are proud of having science information that has significance in the world that exists beyond the classroom. For this reason, it is important to bring visitors engaged in or knowledgeable about the sciences to the classroom if school policies limit field trip possibilities. Classroom windows should not be overlooked as "field" resources for verifying the reality of classroom science learnings. Some teachers consider the street repair crew outside or the squirrel on the window ledge to be distractions that must be resisted. Other teachers consider such occurrences handy and welcome illustrations of the world about which the children are learning.

This holistic approach weaves physical, sensory, and emotional activities into the total intellectual process by encouraging learning in both rational and intuitive styles. This approach is also consistent with the recommendations of the National Center for Improving Science Education, (Bybee, et al., 1989). They suggest six criteria for developing themes and topics for science:

- They build upon children's prior experiences and knowledge.
- They capture children's interest.
- They are interdisciplinary, so that children see that reading, writing, mathematics, and other curricular areas are part of science and technology.
- They integrate several science disciplines.
- They are vehicles for teaching major organizing concepts, attitudes, and skills.
- They utilize these instructional characteristics appropriate for kindergarten through eighth grade science:
  - Each is a hands-on activity, and students' lives and their world are the focus of the activities.
  - Scientific and technological concepts and skills are developed within personal and social contexts.
  - The students construct their concepts, attitudes, and skills through a variety of experiences.

Secondary Integrating Activities

Two important secondary methods of integrating learnings are less direct, since they depend upon the teacher's ability to capture the opportune moments to apply them. They are (a) keeping concepts alive by applying them and (b) helping children relate new information to already acquired concepts.

Children can be greatly helped in thinking through such relationships by themselves if the topics to be linked are presented sequentially whenever possible. The understandings from one set of activities can then lead directly or transfer indirectly to those that follow. As topics are developed, some general suggestions will be made for continuing these conceptual concerns and accommodating new information to already established concepts. These secondary integrations surround the primary integration model as depicted in Figure 1–2.

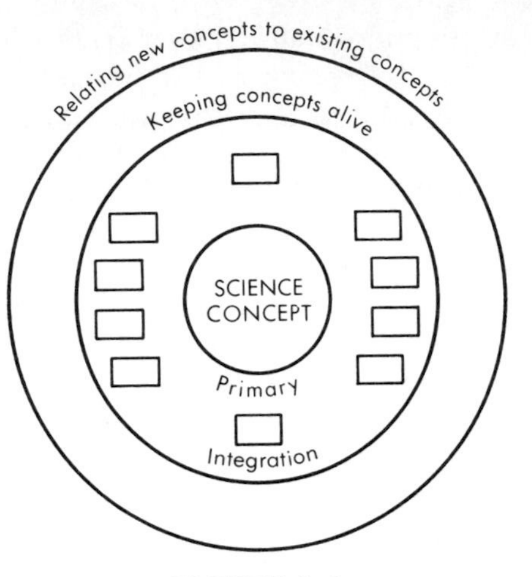

FIGURE 1–2

REFERENCES

AINSWORTH, M. & BELL, S. (1970). Attachment, exploration, and separation illustrated by the behavior of one-year-olds in a strange situation. *Child Development, 41,* 49–67.

BIBER, B. (1979). Thinking and feeling. *Young Children, 35, 4–16.*

BYBEE, R., BUCHWALD, C., CRISSMAN, S., HEIL, D., KUERBIS, P., MATSUMOTO, C., & MCINERNEY, J. (1989). *Science and technology education for the elementary years: Frameworks for curriculum and instruction.* Washington, DC: National Center for Improving Science Education.

CARSON, R. (1960). *A Sense of Wonder.* New York: Harper & Row.

CLARK, B. (1986). *Optimizing learning.* Columbus: Merrill Publishing.

FRAIBERG, S. (1950). *The magic years.* New York: W. W. Norton.

HARLOW, H. (1958). The nature of love. *American Psychologist, 13,* 673–685.

HARMAN, W. & RHEINGOLD, H. (1984). *Higher creativity.* Los Angeles: Jeremy Tarcher.

KAMII, C. & DEVRIES, R. (1978). *Physical knowledge in preschool education: Implications of Piaget's theory.* Englewood Cliffs, NJ: Prentice-Hall.

LANGUIS, M., SANDERS, T. & TIPPS, S. (1980). *Brain and learning: Directions in early childhood education.* Washington, DC: National Association for the Education of Young Children.

RAZRAN, G. (1961). The observable unconscious and the inferable conscious in current Soviet psychophysiology: Interoceptive conditioning, semantic conditioning and the orienting reflex. *Psychological Review, 68,* 126–127.

RESTAK, R. (1979). *The brain: The last frontier.* New York: Doubleday.

SAGAN, C. (1977). *The dragons of Eden: Speculations on the evolution of human intelligence.* New York: Random House.

SELIGMAN, M. (1975). *Helplessness.* San Francisco: W. H. Freeman.

SEVERIDE, R. & PIZZINI, E. (1984). What research says: The role of play in science. *Science and Children, 21,* 58–61.

SPERRY, R. W. (1970). Perception in the absence of the neocortical commissures. *Perception and Its Disorders, 48.*

TIPPS, S. (1982). Making better guesses: A goal in early childhood science. *School Science and Mathematics, 82,* 29–37.

WHITE, R. (1968). Motivation reconsidered: The concept of competence. In Millicent Almy (Ed.), *Early childhood play: Readings related to cognition and motivation.* New York: Associated Educational Services.

2

Science Participants: Children, Teachers, and Families

YOUNG CHILDREN AS THINKERS

If teaching is to be effective, it must be within the range and form of children's ability to grasp meaning from it. For almost two decades, the description of the thinking abilities of young children has been based on the highly respected cognitive development theory of Jean Piaget. Indeed, the rationale for establishing the early childhood period as a separate educational domain evolved from a major Piagetian insight: that children up to the age of 8 need to be actively involved with real materials to acquire and expand their intelligence.

Many of Piaget's theories influence current educational practices. However, inventive studies by contemporary researchers continue to provide new information about how young children develop as thinkers. The data from these studies can enlarge our picture of the intellectual functioning of young children in the preoperational stage of thought. These findings merit consideration in developing a science teaching philosophy, since they shed light on how young children organize their information about the world. Examining Piagetian insights together with the newer findings shows no disrespect for Piaget's pioneering. It acknowledges that the frontier is constantly changing in the dynamic field of cognitive research.

Logic Structures

Piaget's basic tenet was that children build up knowledge actively for themselves as they organize and interpret their experiences. In this way children form typical structures of logical thinking that differ for each stage of cognitive development. Piaget maintained that the progression from one stage of thinking and understanding to another could not be hastened by direct teaching of more mature logic.

Many studies have confirmed that it is futile to verbally teach young children the mature logic of conservation. This is the concept explaining that the quantity of a material or a group of objects changes only by adding to or taking from that quantity, not by rearranging the material or objects. However, young children have been taught to solve conservation tasks by researchers who did not rely on verbal teaching approaches (Siegler, 1978). Instead, they allowed the children to become familiar with the test objects and taught them the skills needed to focus on the important parts of the problem. The trained children were then able to recall and apply the skills to solve those and other kinds of conservation problems.

Rochel Gelman (1969) was one of the first researchers to design such a training study. She found that 3- and 4-year-olds could correctly state whether or not the quantity of items in a conservation task had been changed. These children were not misled by the appearance of rearranged objects, as Piaget believed they would be.

Consistent Application of Logic

Piaget proposed that the general logical understanding that children develop during a given age-linked stage could be applied consistently to a wide variety of similar problems. However, a number of studies examining children's ability to use the logic of conservation showed great variation in the ages and cognitive stages at which that logic could be correctly applied to different types of conservation problems.

Piaget acknowledged these inconsistent findings in his later writing, but he did not attempt to explain them (Fischer, 1980).

Assessment Methods

Many investigators now feel that Piaget's methods of assessing children's intellectual development did not test all of their capabilities. They point out that the incomplete and rather negative description of children's thinking in the preoperational stage was, in part, the result of inconsistencies in Piaget's assessment method.

For example, Piaget identified the infant's capacity to recognize cause and effect relationships during the sensorimotor stage of thought. But he concluded that children in the following preoperational stage were unable to do accurate causal thinking. He believed that children of this general age were too strongly influenced by their personal viewpoints to be able to think logically about the cause and effect of physical events.

Piaget used two different kinds of assessments to reach these differing conclusions. He determined infants' awareness of cause and effect in the way they manipulated objects. But he did not assess preoperational children with objects they could manipulate or even see at close range. Instead, he asked these children to reply to questions about the causes of natural events that occurred far from the children's immediate experience: events such as the movement of clouds, the sun, and the moon. Young children were unable to give accurate reasons for these events. Instead, they used fantasy, the supernatural, and magical explanations.

When Michael Berzansky (1971) investigated young children's ability to do cause and effect thinking, he used a method that was more consistent with the way Piaget tested infants. He found that preoperational children *were* able to do causal thinking about familiar events and about objects they could see and touch, but they were unable to do so when asked about unfamiliar, far-removed events taking place in the sky.

Rochel Gelman developed another consistent method to reveal children's causal thinking. She designed tasks that could be completed nonverbally, instead of requiring verbal responses as did Piaget. She found 3- and 4-year-olds were capable of arranging groups of pictures in correct order to show the causes of certain events. Many of those children were also able to rearrange the cards correctly when asked to restore the original events. This called for flexibility in their thinking, a quality that Piaget found lacking in children of this age (Gelman, 1979).

Robbie Case (1986) took still another approach to getting more consistent information about how children's causal thinking develops over time. He used a single type of equipment, the balance scale, to assess children from infancy to the age of 12. By presenting balance problems of gradually increasing difficulty, he was able to show more precisely a steady progression in children's causal thinking ability. Some of his results will be shown later in this chapter. (See Table 2–1.)

Reversibility of Thinking

One of Case's balance scale problems required children from 3½ to 5 years of age to reverse the thinking they used to succeed with one task in order to succeed with a second task (see Table 2–2). In other words, children had to be able to understand the reversibility of relationships to perform the second task. Case found that children were able to demonstrate this understanding by the age of 5 years. This finding is interesting since Piaget considered the attainment of reversible thinking a signal of the end of preoperational thought. According to Piaget's work, reversibility was generally attained around the age of 7 or 8 years.

Alternative Cognition Theories

There is a key theoretical difference underlying the research methods used by Piaget and those of many of the contemporary researchers. Piaget contended that children could only understand ideas they had developed through their own physical and mental actions. Researchers like Case believe that children do not learn solely by solving problems and exploring on their own. They also learn by imitating and interacting with others. These are the strategies children use from infancy onward to acquire language, relate to others, and learn about the physical world.

When he assessed children's logical thinking, Piaget offered them no explanations about the questions he asked. Many of the newer researchers, on the other hand, allowed children to explore materials, to observe others working with them, and hear others talk about the problems they were presenting to the children. They believed that children best demonstrate their reasoning ability after they have been exposed to the materials and the rules underlying the tasks.

Piaget viewed the development of thinking primarily as an internal process accomplished independently by the child. His structural theory held that broad stages of logical patterns of thinking (structures) emerged slowly from smaller patterns (schemas). He contended that the same broad patterns of logical thought available at a particular stage of development are then used by the child for logical problems in general.

The *information-processing* concept of cognition forms the basis for much of the newer research. In this view, thinking functions and expands steadily as information is acquired and processed in computer fashion. Stored information is used, or processed, with increasing efficiency in the form of specific strategies for solving specific problems. These strategies are gradually consolidated into more complex strategies in two ways: (a) as the developing nerve structures in the brain enable more storage and processing, and (b) as children have opportunities to explore, solve problems, imitate others, and interact with others (including parents, teachers, and researchers.)

The socio-instructional theory of cognition is another current view of the psychology of learning (Belmont, 1989). It is based on the 50-year-old work of Lev Vygotsky. He believed that children's thinking depends upon their level of speech, and it develops as they interact with others. According to Vygotsky, the child acquires new thinking strategies and skills under the guidance of a person who already has that strategy or skill: a commonsense observation that has been validated through research.

A Blended Theory

Case has blended the strengths of both the Piagetian and information-processing approaches into a single, comprehensive theory. In his book *Intellectual Development: Birth to Adulthood,* Case presents the child as a problem solver who develops increasingly complex thinking strategies to handle all aspects of

TABLE 2–1 Spatial reasoning strategies at various ages: Colored shapes classification

$3\frac{1}{2}$ to 5	5 to 7	7 to 9
One-dimension classifying strategy used: color *or* shape ("the blue ones" *or* "the round ones").	Coordinates classifying strategies. Uses two dimensions: color and shape ("red squares"). Coordinates one-dimension classifying strategy with counting strategy ("more red ones than blue ones").	Coordinates two-dimensional classifying strategy with counting strategy ("more round ones than white ones").

Data from *Intellectual Development: Birth to Adulthood* by R. Case, 1986, New York: Academic.

life. Case incorporated into his theory the existing developmental information on children's social interactions, language development, and spatial activities (block-building, drawing). He has shown that developmental steps to greater levels of complexity in those areas of life directly parallel the increasing complexity in children's mathematical and scientific (cause and effect) reasoning (Case, 1986).

One such parallel can be seen by comparing the results of two studies of young children's thinking strategies. Tables 2–1 and 2–2 show strategies used at different ages for a spatial reasoning task (classifying colored shapes), and for a cause and effect reasoning task (predicting effects of weight placement on a balance scale). Case's research indicates that when young children can use their natural tendencies to actively explore and solve problems as well as to imitate and interact with others, the following thinking abilities may be shown:

1. Continuous ability from infancy onward to do cause-and-effect (scientific) reasoning.
2. Ability by age 5 to do reversible thinking.
3. Ability by age 7 to focus on two parts (dimensions) of a problem at one time, *and* to use information from one part to reason about the other.

While information-processing research may still provoke controversy, it can help us become more aware of the problem-solving skills that children use spontaneously in natural life situations. It can help us respect the intricate mental processes required for a child to attain the seemingly simple goal of getting a toy from a shelf beyond reach. The processes needed to solve this problem require sizing up the pertinent facts of the situation, organizing the information, making inferences, and planning and carrying out the solution successfully (Klahr, 1978).

TABLE 2–2 Cause/effect reasoning strategies at various ages: Balance scale problems

$3\frac{1}{2}$ to 5	5 to 7	7 to 9
One perceptual strategy used to predict side of balance that will go down ("Side with heaviest looking stack of weights will go down."). By age 5, strategy can be reversed when asked in a later problem ("Which side will go up?").	New strategy integrates use of counting strategy to reason about another dimension ("There are more weights in this stack, so it weighs more, and this side of the balance will go down.").	Integrates weight and count strategy with reasoning about the new dimension of distance from fulcrum ("Stack farthest away will go down") if both stacks are of equal weight and number.

Data from *Intellectual Development: Birth to Adulthood* by R. Case, 1986, New York: Academic.

The implications for teaching suggested by these studies must be weighed carefully. We must distinguish between what is possible for children to do and what, in long-term consequences, is desirable for them to do. The findings do not imply that earlier demands should be placed on young children to learn specific science content. They do, however, reveal beginning capabilities that can be encouraged through support for curiosity and problem-solving opportunities. They explain how young children are able to reason about and recall relevant science experiences that they choose to pursue.

YOUNG CHILDREN AS PERSONS

Earlier we looked at the influence children's feelings about themselves and their physical world have on their ability to think and learn. Learning, or failing to learn, is also strongly influenced by children's feelings about their place in the social world. The characteristic ways children feel about themselves and relate to others are central parts of the personality. Erik Erikson (1977) identified general developmental trends in personality growth that both contribute to and are enhanced by successful learning. They are significant affective components of cognitive development.

Sense of Initiative

If children's earliest experiences have promoted trust in others, and if they have then developed a good sense of autonomy, the next positive personality trend, initiative, unfolds. Three- and 4-year-olds who are developing a sense of initiative have great energy to invest in activity. They want to know what they can do. They are eager for new experiences and new information about their world. Emerging reasoning powers blend with a developing imagination, leading the child to ask searching questions. Children's delightful curiosity in this phase can be, and endlessly is, expressed as "Why?" A budding behavior control system allows the child to slow activity long enough to become immersed in events that capture attention.

Preschoolers seek the acceptance of children who are like themselves in some way and have mutual interests. This sociability means that a preschool child rarely works alone at the science table, though sharing equipment and materials may still be somewhat difficult. In general, personality developments in this stage make this an ideal time to explore with children the regularities, relationships, and wonders of the world close at hand.

Sense of Industry

Young school-age children who have succeeded in developing a sense of initiative move toward the next positive personality trend that Erikson described as the sense of industry. Eagerness to do meaningful tasks and to become good at them is evident. Children can persist in working at a task until it is completed for the sake of satisfaction in accomplishment. During this phase, children's sense of competence is enhanced by mastering challenges. These children are able to work cooperatively with others because they have good inner control over their behavior. The desire to follow the classroom rules is especially visible in kindergartners. If all goes reasonably well in children's lives, this positive personality trend continues through the remaining early childhood years.

Respecting Personal Development Traits

The natural tendencies to learn where one belongs and what one can do in the social and physical worlds, to create and imagine, to seek answers, to become increasingly effective as an active doer and thinker, and to take pride in independent accomplishment and growing competence all support the interest and persistence needed to learn throughout life. But these tendencies and abilities can be misdirected and even diminished to the point of burnout when they are channeled too early into formal training of academic skills and fact memorization (Elkind, 1987). These tendencies are respected and nurtured when preschools and kindergartens allow children to choose whether or not to participate in science explorations.

THE TEACHERS

Who Can Teach Science

Any teacher who is capable of maintaining a classroom atmosphere of warmth, acceptance, and nurturance has the basic qualifications for guiding young children in discovery science. A positive attitude toward science, and the ability to carry out the enabler, consultant, and facilitator roles are needed for good teaching. However, little can be taught when a close rapport between child and teacher is missing. Children learn most from people with whom they feel the bond of personal interest and caring.

Attitude Contagion

Long-term attitudes toward science as subject matter begin with the attitudes of teachers whom children encounter in their earliest exposures to science. Positive attitudes toward teaching science may have a long history. For some teachers, those good feelings accumulated during their own satisfying exposures in elementary and secondary schooling. Other teachers retained strong science interests in spite of inadequate school experience. They may have had an "answer person" in their lives: a parent, a grandparent, or perhaps a camp naturalist, who had interesting information to share and the patience to help a child find answers. People with such fortunate backgrounds feel comfortable teaching science, and they recognize the importance of science in their students' lives.

Children's negative attitudes toward science may have been "caught" from unenthusiastic teachers who relied on textbooks as the only source of gaining information. Those teachers may have been too unsure of themselves to allow children to consider alternatives to the textbook explanations, or to find out through experimentation. Many female students lost interest in science when they came up against gender bias in the classroom. They observed male students receiving preferential teacher attention because of the stereotyped assumption that males have a stronger aptitude for science (Sprung, 1985). Still others have been discouraged by science courses that emphasized memorizing facts rather than understanding the principles that affect our daily lives.

Negative attitudes toward teaching science can be turned around by opportunities to observe and participate in hands-on science experiences with eager young children. Participation in activity-based science education courses or workshops can restore lost confidence and revitalize faded interests. Classroom follow-up studies of teachers who have participated in activity-based science workshops reflect shifts toward more positive attitudes. The majority of the 400 teachers who were observed as they taught were found to have greatly improved their skills and to have increased the amount of class time they devoted to science activities (Bredderman, 1982).

Authentic Interest

The teacher's authentic interest in finding out is a vital part of the positive teaching attitude. This means a willingness to learn along with the children when they lack answers. The ability to admit to not knowing everything was one of the traits of a good teacher, as identified by older students in a cross-cultural study (Hofstein, 1986).

The teacher's own interest in finding out sustains the children's ready curiosity. It can revive curiosity in children who have been belittled for asking questions, and it can rebuild curiosity that has atrophied in an unstimulating environment. When the teacher's own sense of wonder is alive and active, curiosity behavior is modeled for the children. This important attitude has a place in the science-teaching framework, as indicated in Figure 2–1.

Teaching Roles

Effective discovery-science guidance calls for four teaching roles:

1. The *facilitator* creates a learning environment in which each child has a chance to grow. Planning, gathering needed cast-off materials, and actually trying experiments are science facilitator tasks. In this role there is a tolerance for messiness as children work, a willingness to risk new ventures, and the ability to profit from mistakes.
2. The *enabler* turns on children's intellectual power by helping them become aware of themselves as thinkers and problem-solvers. This role contrasts with the "teacher" image so many of us carry from our own school days in which the teacher seemed to be the ultimate source of all knowledge. That teacher could dim children's intellectual power by somehow magnifying the distance between the knowledge in his or her possession and that of the child. The

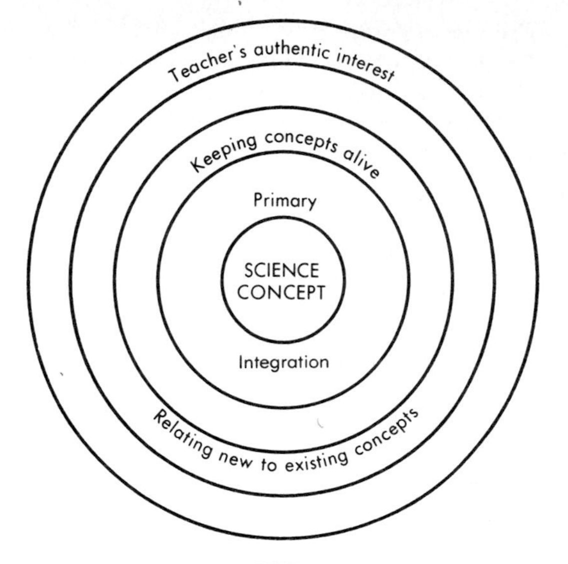

FIGURE 2–1

enabler, on the other hand, sets a positive, encouraging tone by staying in touch with his or her own excitement in discovery.

3. The *consultant* observes carefully, listens closely, and answers questions simply, as children engage in their explorations. In this role, small bits of information can be offered as learning cues, and questions can be asked of the child to help focus on the relevant parts of a problem. The consultant allows the child time to reflect on the new idea and tackle the solution independently. This role often intimidates beginning teachers until they can accept themselves as learners, too. The consultant is a supportive coaching role rather than a directive one.
4. The *model* deliberately demonstrates to children the important traits of successful learners such as curiosity, appreciation, persistence, and creativity. Lillian Katz (1985) defines these characteristics as *dispositions:* habits of mind or tendencies to respond to situations in particular ways. Katz points out that, though these qualities are essential, adults rarely identify and demonstrate these dispositions toward learning. Yet, these dispositions are best learned through example. We model these positive learning dispositions when we share our personal experiences and thought processes with children. We model persistence by describing our own efforts, for example: "At first I couldn't make it work, but I just kept on trying till it did work." We model creativity when we reveal a problem we solved in a new way, for example: "I needed a better way to let you see both sides of the moth. Then, I had an idea that worked. I taped these two plastic lids together to make this neat display case for the moth."

The teacher who holds positive attitudes and understands the necessary roles is well rewarded in both personal growth and in the pleasure of sharing the sense of mastery that children find through their science discoveries.

THE FAMILIES

All parents have information about the world that can be passed on to their children—to their mutual benefit. Jim's dad let him scrub tires and wipe grease off bolts for him as he salvaged parts from the derelict cars that stood in their front yard. As a consequence, Jim was able to make contributions to class discussions about simple machines after weeks of virtually silent presence in school. Daniella's grandmother contributed the cecropia cocoon that, months later, provided us an hour of breathless wonder as the moth struggled free in the full view of 20 children. Five-year-old Tim was able to share solid information about many kinds of planes and how they flew, because his pilot father took him to hangars to see the planes and watch the mechanics at work. These family members enriched the learning of a whole class. They also intensified the interest of their own child in knowing the world by endorsing the value of that knowledge.

The teaching role of the parent (or parent figure) is unique and vital. Parents, not classroom teachers, have the only continuous opportunity to guide a child's intellectual growth. The earliest imprinting of attitudes of appreciation or disregard for the natural or man-made world takes place within that primary relationship. This teaching by example occurs as part of the child's socialization, whether consciously or not.

The power of parents' supportive responses as motivation for children's immediate learning achievement has long been documented (Crandall, 1965). More recent evidence demonstrates the long-lasting effects of parental encouragement on the degree of interest of women and minorities in college math and science (Thomas, 1986). Bloom's 1985 study of the early development of successful, talented adults (including scientists) consistently revealed parental support for the child's special interest, as well as encouragement to persevere and pursue those interests wholeheartedly. Our own science teaching can help generate parental interest where it might be lacking. When children take home evidence of practical problem-solving ability or share science information that has meaning in the adult world, parents rarely fail to respond positively.

Since its inception, Early Childhood Education has striven for a teaching partnership with parents. Currently, the U.S. Department of Education has set the goal of enlarged parent involvement as one way to revitalize science education. Unfortunately, many parents do not realize that they possess science knowledge that could be comfortably shared with their children. They may never have experienced the companionable pleasure of watching an impressive caterpillar with their child. They may not have paused to take a close look with their child at the fascinating gears on a bank vault door.

Mother lifts Maya for a closer look at dozens of gleaming gears. She validates and rewards Maya's curiosity.

To help parents promote children's interest in science, informal, enjoyable *Exploring at Home* activities have been described for each of the major science topics presented in Part Two of this book. These projects are intentionally presented in a lighthearted way to avoid the inappropriate performance demands by parents that discourage rather than encourage children's efforts. Each of these related activities may be duplicated by classroom teachers and sent home with the clear intent of offering family fun, rather than required homework. To emphasize the special role the family plays in children's education, most of these activities are those that are best done at home at night, that directly relate to the home, or that are long-term, beginning-to-end cycles that can have special value for families (Furman, 1990). A bibliography listing home-based science project books of special interest to parents appears in Appendix 2. The influence of parental support completes the science learning framework, as shown in Figure 2–2.

A growing number of cities now have children's museums featuring interactive, physical and natural science discovery areas. Many are specifically designed for family involvement. Others, like the Fort Worth Museum of Science and History, include a full schedule of informal classes from preschool through high school. Teachers can inform parents about the nearest science museum or nature center with exhibits for children. A free directory of science museums is

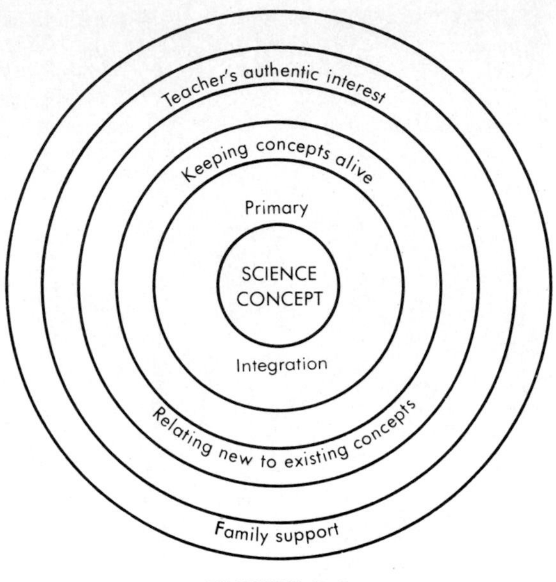

FIGURE 2–2

available from the Association of Science Technology Centers, 1413 K Street, NW, Washington, DC 20005.

REFERENCES

BELMONT, J. (1989). Cognitive strategies and strategic learning. *American Psychologist, 44,* 143–148.

BERZANSKY, M. (1971). The role of familiarity in children's explorations of physical causality. *Child Development, 42,* 705–715.

BLOOM, B. S. (Ed.). (1985). *Developing talent in young people.* New York: Ballantine Books.

BREDDERMAN, T. (1982). What research says: Activity science—The evidence shows it matters. *Science and Children, 20,* 39–41.

CASE, R. (1986). *Intellectual development: Birth to adulthood.* New York: Academic.

CRANDALL, V. (1965). Achievement behavior in young children. In W. W. Hartup and N. L. Smothergill (Eds.), *The young child: Reviews of research* (Vol. 2, pp. 165–185). Washington, DC: National Association for the Education of Young Children.

ELKIND, D. (1987). Superbaby syndrome can lead to elementary school burnout. *Young Children, 42,* 14.

ERIKSON, E. (1977). *Childhood and society.* New York: W.W. Norton.

FISCHER, K. (1980). A theory of cognitive development: The control of hierarchies of skill. *Psychological Review, 87,* 477–531.

FURMAN, E. (1990). Plant a potato—Learn about life (and death). *Young Children, 46,* 15–20.

GELMAN, R. (1969). Conservation acquisition: A problem of learning to attend to relevant attributes. *Journal of Experimental Child Psychology, 7,* 167–187.

HOFSTEIN, A., SCHERTZ, Z., & YEAGER, R. (1986). What students say about science teaching, science teachers, and science classes in Israel and the U.S. *Science Education, 70,* 21–30.

KATZ, L. (1985). Dispositions in early childhood education. *ERIC/EECE Bulletin, 18.*

Klahr, D. (1978). Goal formation, planning and learning by preschool problem solvers. In Robert Siegler (Ed.), *Children's thinking: What develops?* Hillsdale, NJ: Earlbaum.
Siegler, R. (Ed.). (1978). *Children's thinking: What develops?* Hillsdale, NJ: Earlbaum.
Sprung, B., Froschl, M., & Campbell, P. (1985). *What will happen if . . . Young children and the scientific method.* New York: Educational Equity Concepts.
Thomas, G. (1986). Cultivating the interest of women and minorities in high school mathematics and science. *Science Education, 70,* 31–34.

3

Guiding Discovery Science

DISCOVERY SCIENCE

In discovery science, learners actively seek information. Regardless of the level of sophistication, be it science in the research laboratory or in the preschool, science is a way of thinking and gaining knowledge that consists of four steps.

1. Becoming aware of a problem *(noticing)*
2. Hypothesizing, proposing an explanation *(wondering why)*
3. Experimenting *(finding out)*
4. Communicating the results *(talking about it)*

Science activity can also be described in terms of the methods or processes for carrying out scientific activity: observing, classifying, measuring, communicating, experimenting, predicting, inferring, designing investigations, interpreting, and constructing explanations from the data. The least complex of these processes, observing and rudimentary experimenting, can be seen in the behavior of infants. As their thinking develops toward the final stages of abstract thought, children can use the processes with increasing sophistication.

Some science educators advocate basing science programs for primary and elementary grades entirely on the development of skill in using these processes without focusing on specific science content. Others feel that to separate the learning process from that which can be learned denies children the opportunity to satisfy curiosity and to acquire helpful information. It is quite possible for children to become proficient at using the science processes, while at the same time gaining concepts through hands-on discovery activities.

One of the best means of helping children know the world at hand is to organize materials so they can explore, question, reason, and discover answers through their own physical and mental activity. This guided discovery approach to science learning emphasizes *how* to find answers, as well as *what* can be learned. When discovery science experiences are seen as part of the child's continuous search for knowledge, it makes good sense to support and enable that search in the classroom in several basic ways:

- Discovery science values and rewards curiosity as a valid learning tool (Kyle, Bonnstetter, McCloskey & Fults, 1985).
- Discovery science encourages individuality and creativity in children's problem solving. This leads to better retention of the concepts attained. Interestingly, the four steps of science as a way of knowing and thinking are also those used to describe the creative thinking process.
- Discovery science validates the fundamental learning style of direct involvement with materials. It builds upon spontaneous experiencing, touching, and trying by the very young. Active participation remains so important to knowing throughout life that our language uses the term *firsthand* to mean something that is known from the original source.
- Discovery science experiences in early schooling help to gradually replace the young child's intuitive explanations of the unknown. They demystify bewildering events, but respectfully retain the profound and beautiful aspects of natural occurrences.
- Discovery science provides a means of focusing the attention of the restless child, the anxious child who is preoccupied with personal concerns, and the bored child who is understimulated by less challenging aspects of the curriculum.
- Discovery science has appeal for the resistant learner, the child who has come to protect herself from inappropriate pressure to perform formal learning tasks. The play-like quality of exploring real materials reduces the stress that more precise "paperwork" tasks produce. It allows more physical and social involvement than do the more structured forms of schoolwork. Pressured children can put aside their resistance and restore their sense of themselves as learners when they engage in science activities.
- Discovery science, by its very nature, provides an intriguing path to the goal of developing children's intellectual potential. Any form of learning that includes the manipulation of interesting materials is highly motivating to children. The discovery science activities and cross-curricular extensions in this book lend themselves to programs where language learning and content are integrated, such as the Language Experience Approach, the Whole Language Approach to reading, and the Project Approach to curriculum development. The sensory and psychomotor extensions fit well into English as a Second Language programs that emphasize total physical response as a learning strategy.

GUIDING LEARNERS

Teaching Styles

Opinions differ about the teacher's role in experiential learning. Some educators interpret Piaget's statement, "telling is not teaching," quite literally. They consider that any verbal instruction from a teacher interferes with the child's genuine understanding of new ideas.

Other educators and researchers assign a more active role to teachers in the child's acquisition of knowledge. They believe that children do benefit from appropriate forms of verbal guidance. Robbie Case (1986) points out that a teacher who can empathize with a child's efforts and share a child's feeling of excitement and satisfaction in solving problems is better able to guide and encourage children's intellectual development. Case believes that children learn by imitating and interacting with others, as well as through their own exploration and problem solving. He sees the teaching role as social facilitation of children's problem-solving abilities.

Guided discovery learning calls for this social facilitation as direct and indirect guidance. The teacher supplements the child's active exploration of problems by helping the child draw meaning from the experience and by extending conceptual learning. The affective component is also important in this teaching/learning process.

Organizing Approaches

There are two distinctive approaches to organizing socially facilitated science activities: child-instigated experiences (the incidental approach), and teacher-instigated experiences. The child-instigated approach is ideal for very small classes of preschool children, and for nurturing the individual interests of gifted older children.

Incidental science can occur at any time or place, whenever a child's curiosity is aroused by something significant: a sparkling bit of quartz embedded in the sidewalk, the iridescent sheen of a beetle's back, or an evaporating playground puddle. The teacher capitalizes on the child's discovery by asking questions that lead to further discovery, by relating the find to something the child already knows, by extending the experience into other classroom activities, and by offering to help the child locate other resources for expanding his or her information. The confident teacher with a ready interest and strong background in science can do an admirable job of enlarging children's information this way.

There are drawbacks to this approach, however. It is difficult to provide the same degree of support for larger classes, or to fit spur-of-the-moment activities into a more structured school program. Nor do these incidental, child-instigated investigations constitute an adequate conception of science. Science is not only a collection of facts about the physical world, but also a way of describing and explaining physical phenomena, and organizing them according to relationships among them.

The teacher-instigated approach outlined in this book centers on interests common to young children. It places related experiences within a framework of corresponding extensions. The science experiences themselves are not intended for random presentation as interesting time-fillers. Rather, the experiences and suggested extensions are presented in a unified way, perhaps continuing for a week or two. It is tempting to regard units of learning as accomplished work that can be neatly tied up and shelved for the year. However, lasting learning takes place through many encounters with ideas in a variety of contexts. The learnings can be referred to and renewed throughout the school year when they relate to other learnings and classroom occurrences. Recent research by Bahrick demonstrated that material reviewed and built onto over a period of time can be retained remarkably longer than material that has only been presented once (Adler, 1991). Science knowledge further takes hold when it has utility, and when it becomes part of a continuous chain of learning that develops throughout life.

PREPARATION FOR TEACHING EARLY SCIENCE

There is no question among educators about the superiority of active discovery over the textbook and lecture method of learning science. More than 300 comparative performance studies in recent years have confirmed that superiority (Shymansky, 1984). Yet it appears most kindergarten and primary grade teachers continue to use the less effective methods. The most common reason teachers give for not offering hands-on learning activities is their conviction that they lack enough science background to be able to answer children's questions. The open-endedness of problem solving bothers them. They feel uneasy when more than one solution can be appropriate. When these science anxieties go unchecked, they can be easily transformed into a list of airtight reasons for reading about, instead of "doing," science.

A sense of inadequacy about science can be a real limitation to science teaching because it tends to be communicated nonverbally to children as distaste for the subject matter. However, it is equally possible to do poorly at teaching youngsters because of being over-trained in science! It can be difficult for sophisticated scientists to translate their verbally elaborate concepts to a level of ideas that young children can understand.

Actually, most of us have much more practical science information than we may recognize. We can fill in gaps in our knowledge as we prepare to teach. A beginning teacher can supplement basic knowledge by carefully reading the *concepts, learning objective statements*, and *activity directions* in this book. Trying out the experiments and recording the results, problems, and personal reactions to the experience can increase a novice's confidence. Reading some of the starred reference books, written at the young child's level of understanding, will add appropriately to one's knowledge base. Reference books suggested for teachers will provide depth of information to enlarge the teacher's developing interests and provide answers prompted by a child's urgent curiosity.

The recommended children's reference books and science-based stories for children, especially those by Franklyn Branley and Milicent Selsam, are written by authors who know both the science concepts and young learner's capacity to absorb core ideas. Many of these fine writers also know how to incorporate humor and excitement into their work.

There is merit in renewing and strengthening our science information just before we teach. Then the satisfaction gained in becoming reacquainted with the concepts will be fresh and available to spark the interests of the children. Finally, it is reassuring to remember these words by Jerome Bruner (1961): "The basic ideas that lie at the heart of all science are as *simple* as they are *powerful*."

INDIRECT AND DIRECT TEACHING

Discovery learning is guided *indirectly* through thoughtful questioning and listening, and by sensitive discussion-leading. It is guided *directly* by offering conceptual cues and by encouraging effort. Since many of us were not exposed to these kinds of leadership in our own schooling, it will be helpful to analyze these techniques.

Thoughtful listening to children's ideas is an indirect way to guide and sustain their interest in discovery learning. We do this when we avoid passing judgment on a child's erroneous or fanciful ideas that differ from a scientific explanation. Susan Carey's research (1986) confirmed that preschool children can have already formed naive, intuitive explanations of biology, based upon their own interpretation of experiences. There are ample anecdotes from teachers and parents to suggest that the same is also true of other topics that intrigue children. When there are gaps in the information they can assemble, children typically fill in those gaps imaginatively to form their personal theories, as did Matthew (Chapter 1). This process is not unique to children, however. Countless adults still believe naively from the evidence of their own perceptions that the sun orbits the earth. Each of us may be able to recall some of our own "childish" misperceptions; ideas that we perhaps dared not express for fear of being laughed at. Had we felt it was safe to reveal them, those misperceptions might have been cleared up much sooner. Children's confidence is sustained when we respect these naive attempts to figure things out instead of devaluing them as amusing, quaint, or as serious misconceptions to be eradicated immediately. The intuitive theories can be gradually clarified and reorganized when discovery experiences reveal the more systematic and complete way of looking at things that we call science.

We indirectly guide children's thinking when we show them our own reasoning processes by "thinking out loud." That could occur when we say something like this: "When I found this tiny leaf on the sidewalk, I thought to myself, 'This isn't like any other tiny leaf I've seen before—Hmmm. But it has five points, like a star. A big sweetgum tree leaf has that shape, too. I wonder if a new sweetgum leaf starts out as tiny as this.' . . . So, what do you think I did to find out for sure?" Sylvia Farnham-Diggory (1990) describes this important modeling as making invisible thinking processes visible to children.

LEARNING TO QUESTION

Just as the quality of scientific research depends upon asking worthwhile questions, the quality of discovery science learning is indirectly guided by helpful questioning. Open-ended questions are useful for generating several appropriate answers (divergence). *Divergent questions* can serve many purposes:

- *Instigating Discovery:* A science activity becomes a discovery challenge when it is initiated as a question to answer. Each activity suggested in the following chapters is headed in bold type by the question it can answer. These questions can be used to introduce the activity and to elicit children's conclusions about their activity. The questions can be printed on index cards and posted in the science area to help reading children stay focused in their work.
- *Eliciting Predictions:* Before children experiment, elicit their predictions. "What do you think will happen if . . . ?"
- *Probing for Understanding:* "Why do you think that side of the balance went down?"
- *Promoting Reasoning:* "Why do you think this arm feels dry, and this arm feels wet?"
- *Serving as a Catalyst:* Sometimes a question can be a catalyst to spark renewed interest in a problem, whereas a specific direction from the teacher could make finding out unnecessary. "What could you change to try to make your lever work?" encourages new effort. "If you move the block closer to Robert, you'll be able to lift him with your lever" discourages independent effort.
- *Encouraging Creative Thinking:* In a group discussion ask questions such as, "How would our life be different if there weren't any friction?"
- *Reflecting on Feelings:* In a group discussion, ask questions such as, "What was the best part of the dark box experiment?"

It is also suitable to ask certain *convergent questions,* those with a single answer, to stimulate thinking and exploring:

- *Directing Attention:* By asking a question like "Does the red cup hold as much water as the blue cup?" we can direct a child's attention to a key part of the activity she has overlooked. The child can then correct her own course of action without feeling criticized.
- *Recalling the Temporal Order:* "What did you do first? What did you do next?" "When did this happen? What happened afterward?"
- *Recalling Prior Conditions:* "Does the jar of beans look just the same today as it did yesterday?"

Follow-up questions to convergent questions can be divergent. They can lead to further reflection or to fresh experimentation: "How can you find out?"

It is helpful for beginning questioners to role-play with other beginners in order to improve and feel at ease with new techniques. It can take concerted effort to change the way we shape our questions to overcome habitual closed questioning. Many of us use the pattern of stating the answers we hear from children in the form of a question. We may say, "So the cup of snow melted into a smaller amount of water, right? Isn't that what you found?" This style of questioning reduces children's need to discover answers for themselves, or tells them that the main discovery is to find out what the teacher wants them to say.

LEADING DISCUSSIONS

Discussions serve different purposes at different points in learning new concepts:

- *Introductory discussions* whet interest in a new topic when children are encouraged to recall events they have encountered personally and to contribute what they already know about the subject. At this time children are encouraged to raise their own questions about what they would like to find out in the activities.
- *Small group discussions* follow the children's individual activity to process what took place in their experimentation and to help clarify their thinking. Those discussions can reveal that more than one conclusion can be reached from an activity.
- *Summary discussions* with the whole class pull together the concepts that have been explored in the activities and extensions. Each of the following chapters offers suggestions appropriate to summary discussions, under the heading *Secondary Integrations.*

Group discussions allow children to learn from one another, *if* the teacher models respect for the ideas and experiences that children express. Quiet children may have to be drawn out in low-risk ways: recognize a nod of agreement or a responsive smile from such a child as involvement. If some children have trouble staying involved in large group discussions, this concrete illustration can be helpful:

> Ask each of two children to get a crayon and exchange it with the other child. Have them verify that each had one crayon *before* sharing with the other, and each still has only one crayon *after* sharing with the other. Next, ask each child to share an idea with the other on a topic such as the day's weather. After each has shared an idea, point out that now each child has two ideas about the day's weather. Add "We grow in ideas when we share and listen to each other."

It is important not to close off the exchange of ideas as soon as a child offers key information. Keep the discussion open until each willing child has had a chance to be heard, even if the ideas begin to echo one another.

Adults often overlook the fact that it takes a little time for children to produce thoughtful answers to questions. Many people feel that they are unsuccessful at teaching if their questions are not answered immediately. The results of Mary Budd Rowe's (1974) research indicate that just the opposite is true. She found that the teachers she observed gave children an average of under one second to answer questions. Teachers who were then trained to wait three seconds or more for responses elicited a greater number of answers, longer answers, and more varied and accurate responses. Furthermore, as the teachers waited longer, children began to listen to and respond to each other's comments. Children other than the "brightest" in the class began to contribute answers also. This suggests that when quick responses by a few children end a group discussion, many untapped ideas may be cut off. The practice leaves many children feeling less confident about themselves as thinkers and learners.

A good discussion puts the teacher in the enabler role by empowering children's ability to think and express their ideas. This seems to be a difficult shift for some teachers to make, especially if they are accustomed to having their voices dominate the classroom. With practice, it becomes easier to support and elicit the children's comments using bridging remarks like, "That was a good idea, *and,* others may have different thoughts about it. . . ."

The enabling teacher underscores what children accurately contribute, adds bits of information to expand those ideas, and clarifies misconceptions that might still linger. The teacher summarizes the various points children made in the discussion. If the contributed ideas do not include certain salient points, the teacher can add "Scientists also tell us that . . ." Group discussions are most successful when they are guided with the goal of stimulating the children's thinking and reasoning power.

ORGANIZING TIME AND SPACE FOR SCIENCE

The logistics of making exploration time available to small work groups will vary for each classroom. Now that learning centers are more commonly used, teachers are becoming adept at flexible scheduling of activities. Problems of managing science activities can be eased with the help of an assistant in the room, whether it is a volunteer parent, a retiree, or an "exchange student" from an upper grade. It is best to delegate overseeing other ongoing activities to the helper. In some instances, simple directions typed on index cards or tape-recorded directions can allow children to handle science projects fairly independently.

A good location for science activities facilitates thinking by inviting children to participate and by controlling distractions. Proximity to storage and cleanup aids is important. Varying the setting for activities can build interest. Anticipation is heightened on the days when science takes place under a blanket-covered table!

Elaborate bulletin boards aren't needed to attract children's attention to science when a frequently changed, PLEASE TOUCH display shelf is available. Many teachers begin the school year with noble intentions of welcoming nature finds and other objects of science interest that children bring from home. Perhaps they initiate the project attractively with a bird's nest propped in the crotch of a small tree branch, some special rocks, and a recently shed snake skin. If the goal of a changing display is forgotten, the old things will lose their meaning. Since there is little appeal in a dusty nest or a tattered snake skin, it is better to retire the too-familiar objects. An easy way to keep the science display pertinent to ongoing activities is to try to leave out indestructable materials being used in the discovery activities.

INTRODUCING SCIENCE ACTIVITIES

There are many creative ways to focus children's attention on a new science activity. A spur-of-the-moment thinking game could build interest: "There's something that we see in our room every day that we will use in science today. It has a handle on the outside and rollers on the inside." One relaxed teacher lures his students as he patiently searches through his pockets for the "something I thought you would like to know about." Opening a brown bag to reveal an unexplained object of science interest can provide enticement to find out more. Any of these attention hooks need to be followed with, "What do you know about this?"

A topic can be introduced by reading a related story or poem, and following it with an open-ended question to be answered by the discovery activity. For example, the topic of light could be introduced by reading *Bedtime for Frances* and then asking, "Why do you think Frances was confused and scared by her robe on the chair?" A simple description of an occurrence could start the explorations. "Today, when I stood on my porch, a heavy, noisy repair truck rumbled by and I felt the porch shake. Have you ever felt shaking near something noisy? . . . Today you can find out more about shaking and sounds happening together." While it is not advisable to oversell science as fun and games, with a little care it is possible to avoid making science seem an irrelevant chore.

GUIDING EXPLORATIONS

"Hands-on" for children should imply, as much as possible, "hands-off" for teachers. There will be occasions when a tactful offer to steady a screwdriver or to knot a parachute string can help a discouraged child achieve success. But unsolicited help should not be given in the interests of saving time.

The consultant role is a delicate one: quietly offering a reasoning cue to highlight the important part of an experiment, or asking a question and allowing the child time to reflect on the answer. (Reasoning cues can be taken from the concept statement headings for activities in this book: "You pushed *down* to lift her *up* with the lever.") Wise consultants resist the temptation to hurry a child along to discover what they, the consultants, already know. They are careful not to smother a child's tender spark of inquiry with a heavy blanket of directions and facts. Experiential learning is weakened when a child is rushed to a premature conclusion or "helped" out of arriving at the child's own solution.

SUPPORTING INTEREST IN KNOWING

It would be unrealistic to assume that the spontaneous desire to know will consistently lead all children in a group to take part in every science activity made available to them. It would certainly abuse the developmental goal of encouraging autonomous decision making to require the whole group to engage in each science event that the teacher presents. Knowledge of the uniqueness of individual development should make it clear that each child brings different capacities and motivation into learning situations.

Subtle directives from families can strongly influence the choices children make in school. Direct or indirect messages like, "Be careful with those good clothes . . . Boys don't cook . . . Girls are afraid of spiders and worms," can be powerful inhibitors to children who are unsure of their affection rating with their parents. The directives may curb a child's interest in trying new experiences.

Children also may have needs that may take precedence over the desire to find out. Children who are lonely at home may be very reluctant to pass up any chance to play with special school friends. They may choose a science activity only when friends make the choice. Children who are tired, hungry, or feeling under par may not be able to summon enough interest to observe what is happening at the science table.

If a child consistently spurns science activities, it would be well to think about possible causes and consider ways to make participation easier. The teacher may simply need to ask for that child's help while organizing materials for the day's science activity. Children usually take a proprietary interest in a project that they have helped to prepare. However, when a science topic becomes a central theme threading through many parts of the school program, even the children who do little active exploring will become aware of the concepts being presented. Teachers influence children's interest and involvement with science positively when they value and reinforce the children's focus on learning. They do this by commending children's efforts to observe and explore with specific words of encouragement about what a child is attempting and accomplishing. A comment such as, "Liz, you really looked carefully to find those tiny, folded leaves on your sprouts," helps the child recognize her own competence as an observer. Liz's desire to participate further will be enhanced by such specific support for what she is learning. Encouragement that validates children's competence promotes a positive disposition toward science. It strengthens their concepts of themselves as learners and contributes toward their interest and continued effort to find out more. When the focus is on learning, children are likely to persist in exploring, even in the face of obstacles, because they are trying to understand something new and become more competent (Dweck, 1986).

On the other hand, teachers who value and reinforce correct performance by their classes may have the opposite effect on children's learning motivation. These teachers use automatic, ambiguous praise like, "That's terrific," or "super scientist at work," to keep children "on task." Unfortunately, excessive, empty praise seems to interfere with children's spontaneous interest in an activity (Morgan, 1984). Their attention is directed toward proper performance rather than toward what they are learning. Eventually such inappropriately motivated youngsters tend to give up on tasks when they encounter obstacles. They may only keep working as long as the teacher is nearby to do the "motivating" for them.

DISPELLING STEREOTYPES

Gender and minority stereotypes about intellectual capabilities perpetuate self-doubt and create needless limitations to the development of this nation's problem-solvers. Teachers' actions and beliefs contribute heavily to the self-attitudes being formed by their students. Self-attitudes, whether negative or positive, are the single most crucial force in shaping what an individual is able to accomplish.

While we may voice our desire to help each child develop her or his intellectual potential, our actions may belie our words. Our actions are based on our values, some of which we hold unconsciously. We may be acting on buried, left-over prejudices about the thinking capacity and "natural" interests of women and minorities that were passed on to us as school children. Those stubborn inheritances need to be brought into our awareness and deliberately discarded before they can be replaced by reality-based information. We will truly be acting on our convictions about the problem-solving potential of each child we teach when:

- Girls and minority children receive as much verbal and social encouragement from us as do Caucasian boys—to answer science questions, contribute to science discussions, and persist at experimenting.
- The first chance to try new activities is not the exclusive right of Caucasian boys.
- Science professionals who visit our classes, or whose pictures are posted on our bulletin boards, are as likely to be female and non-Caucasian as they are white males.

When we communicate in these ways our belief that each child can participate effectively in some form of science experiences, we will be making a valid effort to combat crippling stereotypes.

ADAPTING EXPERIENCES FOR YOUNGER CHILDREN

While the activities in the framework were planned for children 4 through 8 years of age, many of the experiences can be adapted for younger preschool children. Two-year-olds can enjoy simple sensory explorations such as feeling air as they move it with paper fans, spinning pinwheels, or swinging streamers on a breezy day; feeling rock textures and weights; touching ice, then touching the water it melts into; watching, then moving like a goldfish; tasting raw fruits and vegetables that have grown from plants; listening to loud and soft sounds; or gazing through transparent color paddles to see surroundings in a new light.

Three-year-olds might be expected to engage in similar activities, taking in greater detail. They will be able to direct deeper attention to such things as exploring new dimensions with a magnifying glass. This group can enjoy some of the classifying experiences on a beginning level: sorting rocks from objects that are not rocks; things that float from those that do not float; and objects that are attracted by a magnet from objects that are not attracted.

ADAPTING FOR CHILDREN WITH SPECIAL NEEDS

Science is an approach to thinking and behaving that has value for children at any level of motor, behavior, sensory, communication, or mental functioning. Because discovery science calls for collaboration among children, it fosters the personal interactions that special needs children often miss. It can help classmates recognize the limited child as an individual with ideas and talents to contribute. Nobody laughed at Lori once her worm-scouting skills came to light. Her problems with pencils and numbers and letters didn't interfere with her passion for capturing insects. Her generosity in sharing them improved her social relationships. Scott's "slow" right leg and droopy right arm didn't limit his deep involvement in science. Frequently left behind on the playground, Scott was in his social and intellectual element at the science table. With good-humored inventiveness, Scott found his own way to water our plants. He soaked a clean sponge in the pitcher of water that was too heavy for him to manage. Then he squeezed for each plant, a simultaneous drink and sponge bath!

Adaptations of materials can enable success for a child with special needs. A plastic basting syringe is easier to use than a small medicine dropper for a child who finds it difficult to grasp objects. Bamboo toast tongs may help children with fine motor control problems pick up small objects.

Changes in approach are important for children with vision and hearing impairments. Advance information about what is going to take place helps the visually handicapped child deal with new experiences comfortably. Face-to-face communication is necessary with the hearing impaired child. Because a hearing loss is invisible to others, it is easy for us to forget that communications are cut off when that child's back is turned.

An easily distracted child benefits from a simplified, clear arrangement of materials to focus upon. Tangible boundaries, such as individual trays of equipment, can help the child with poor control feel assured of a fair share of things to work with. The teacher's close presence can help ensure an island of calm around these children to help them pursue their work. Tiny pebbles or seeds that could be inserted into ears or nostrils should not be available without supervision to children who lack safety awareness. Time given to planning for successful activities minimizes the time and attention such vulnerable children demand when frustrating tasks lead to disruptive behavior.

A chair-confined child will need to have materials brought into close range. A highlight event might be connecting a child with attention to invest and a grasshopper with a new cage to explore. If obstacles to full participation are too great, assign a record-keeping job or lab assistant role, such as keeping track of materials that might be misplaced.

Make use of available professional consultants for specific ways to maximize the potential of a child with a handicapping condition. Until such help arrives, put yourself imaginatively into the handicapping situation to think of ways to work around the child's limitations and through the child's strengths. Whatever accommodations you make to provide satisfying science experiences can help special needs children grow in ability to cope with the challenge. Information on science activities designed for the visually impaired or physically disabled youngsters is available from the Center for Multisensory Learning, Lawrence Hall of Science, University of California, Berkeley, CA 94720.

Children who face difficulty in mastering developmental tasks deserve every chance to develop confidence and initiative that teachers can provide through science activities. A young teacher, Jane Perkins, did this for her class of learning-disabled children. She was deeply moved by their responsiveness to the science activities. She said, "It was the first time any of these children ever expressed curiosity. I'll never again deny them the chance to wonder why."

LEARNING IN THE CONTEXT OF COOPERATION

Current research on children's thinking and learning regards them as results of both individual development and social interaction. There is also growing recognition in our society of the divisive outcome of emphasizing individual accomplishment and competitiveness in the classroom, while failing to teach children how to work cooperatively. Together, these trends provide impetus to the return to cooperative education. The discovery experiences in this book have always been presented primarily as activities to be shared by small groups, with introductions, supplements, and follow-ups as whole-group experiences. They are readily usable in the cooperative learning context.

Management of discovery activities as cooperative learning tasks will vary according to the nature of the project, the availability of materials in quantity, and the experience of the youngsters with handling independent activities (Johnson, D., Johnson, R., Holubec, & Roy, 1988). Certainly, it makes more sense to have only one group of four children working on a particular gravity experience requiring the use of the one available commercial balance and weights, with other groups working on problems suitable for homemade balance scales. Different groups could be using that time to make entries in their topic record booklets or to record their earlier findings on a class bar graph. Other children could be working on sections of the class mural of a gravity-free fantasy or creating a skit about being astronauts.

Precisely how cooperative activities are implemented is a matter for the individual teacher to decide. Surely it is wiser to make a slow transition to cooperative learning groups from a lockstep approach where each child works independently at the same learning task. Typically, it is realistic to plan a science period where discovery activities are paired with less structured activities that require little supervision. Teachers can then devote maximum attention to facilitating the discovery activities. There is no particular merit in attempting to simultaneously juggle a number of differing discovery activities.

Time must be allowed before the science period ends for cleanup and for processing with the whole class what has occurred in the cooperative groups. To ensure that each child in the class has a regular opportunity to report what his or her group has accomplished, a number-rotation system can be used. Each child in a small group is assigned a number from one to four. Then, instead of asking for a volunteer or choosing a high-visibility child to report, the teacher indicates which number will identify the day's reporters.

AREAS OF INVESTIGATION

The science topics that follow were developed in response to the persistent interests, questions, and concerns of a generation of children in the classroom, at play, and at home. They are basic concepts that have functional value in the world of young children to explain how and why familiar events occur. Some topics were developed to modify or replace the naive, magical explanations that children had formulated for themselves. Topics were selected according to the following criteria:

- relevance to the child's immediate experience
- potential for being understood through simple, hands-on activities
- potential applicability to practical child-level problem solving
- significance in promoting safety and physical or emotional well-being
- availability of low- or no-cost materials for experimentation and observation

Two topics that interest many youngsters, prehistoric animals and astronomy, do not meet these criteria for most early childhood settings. They can be more suitably explored by parents and teachers who can provide for children the resources of a natural history museum, a planetarium or observatory, and the night skies. References for developing these topics are listed in Appendix 3.

Early childhood classes may have opportunities to participate in community environmental projects. Every class can incorporate reusing, recycling, and conserving resources into daily routines (Earthworks Group, 1990). Such involvement allows children to have a valued role in significant matters, but it does not take the place of foundation science experiences. Basic understandings of how matter behaves in the natural and physical worlds give meaning to the more complex interrelationships of environmental issues. In the chapters that follow, many hands-on experiences are suggested that lead to broader understandings. The activities are sequenced to provide foundation concepts for subsequent activities.

Other worthwhile science topics with child appeal can be developed by creative teachers, as long as the conceptual purpose and scientific accuracy are clear. The hands-on activities must lead children to a general truth or law about how matter and living things behave. Related extensions in other curricular areas should be just as enjoyable as the active science experiences, but they, too, must have a conceptual purpose. They must not be vaguely connected, aimless things to do that trivialize the science learning.

A teacher's own fascination with an area of knowledge can be a valid starting point for developing a science topic, because that interest and enthusiasm communicate a positive disposition toward science. The adult range of knowledge must then be stepped down to where the children are in their thinking development and to the context of the child's immediate world. Activities then can be designed around the "up close and personal," tangible aspects of that topic. A teacher's commitment to aerobic exercise and optimal health, for example, can translate into beginning experiences with the awareness of air being pulled into children's nostrils and filling their lungs; their own pulse and beating heart, their own moving muscles. The discovery activities could include exploring how breathing rates change and measuring how long it takes to recover normal breath rates after a sprint. A teacher's strong commitment to protecting the environment translates into beginning, hands-on explorations with reused plastic bottle terrariums and expands onto the playground and into the broader community. Bess-Gene Holt (1989) offers excellent leads to exploring the immediate environment and toward understanding the patterns and harmony in the undisturbed natural world.

The activities for the science topics in this book are described as one way of supplying accurate guidance for children's explorations. They are sequenced from the simplest concepts to the more complex. The activities illustrating the more complex concepts could be made available as options for children who are eager to pursue more challenging ideas. Each teacher is in the best position to make this decision. Children are best served and core concepts are more durably retained when teachers offer only a few topics in depth and with cross-curricular reinforcement each school year than when many topics are touched on lightly.

OBJECTIVES AND EVALUATION

The broad objectives underlying the learning experiences in this book focus as much upon the individual's feelings and approach to learning as they do upon the acquisition of information. Observations of the basic harmony in physical and natural world relationships build feelings of security and confidence in children. Reducing fear of the unknown leads to feelings of mastery and strength. Support for curiosity promotes problem-solving ability and offers an avenue for learning how to learn.

In order to quantify certain learning outcomes, behavioral objectives are still used in many schools. However, what a child really learns from an activity might not be immediately evident in the classroom. That feedback may spill out to a parent at home during a sleepy bedtime conversation. The learnings may be fragmentary until they are consolidated through participation in the integrating activities. Or, perhaps the learning is applied nonverbally weeks after the classroom experience has taken place. The real nature of the teaching/learning process is obscured if we assume that children learn only and precisely what a narrow behavioral objective decrees.

Learning objectives are stated in the following chapters for the teacher's use in making curriculum choices. The objective statements include the science processes used in the activities. *Children are not expected to fully absorb objective concepts in their abstract form.* To avoid detracting from the broad goals of the cognitive/affective approach to science learning, objectives are not stated as behavioral criteria for evaluating the child's performance.

Formal evaluation of children's accomplishments in discovery science is required in many schools. One form of assessment is most consistent with discovery science and is most useful for identifying misunderstandings that need further clarification. It is the child's verbal response to questions asked as the child participates in the hands-on activity. Such questions are, "What did you find out about . . . ?" and "Can you show me what happens when . . . ?" For longer plant, animal, or weather observations, the question "What did you find out about . . . ?" asked at a natural closure point provides a gauge for conceptual learning. Children's responses can be recorded most conveniently on a simple checklist. Ideally the check sheet items should center on the grasp of basic concepts, activities completed, and evidence of interest and satisfaction expressed by the child. Such data also serve as self-evaluation for the teacher about the success and value of the learning experiences.

Use of paper and pencil tests of factual knowledge for older primary-grade children can be a barrier to adequate responses, since the flow of ideas can be blocked by the need to write, spell, and punctuate correctly the written answer. This problem can be avoided to a certain degree if the questions can be answered by reference to simple sketches made by the teacher to amplify the written answer.

The assessment method most appreciated by parents at conference time, however, does not involve convenient check marks on a rating scale. Rather, it consists of sharing your casual, on-the-spot jottings in a pocket notebook about their child's intense involvement in a fascinating activity or insightful contributions to a class discussion. A collection of these annual notebooks also sums up the teacher's significant contributions to the lives of children. They form a quiet record of having provided children lasting satisfaction in mastering new ideas. The notes are tangible evidence of having passed along life-enriching awareness of the world we know and appreciate through science knowledge.

GAINING FROM OUR MISTAKES

Many of the materials and activities in this text have been revised as a result of trial-and-error encounters with the logic of young minds. On one occasion, the children in my class were given matching plastic vials, some capped, some uncapped, to use in a buoyancy experience. Then 5-year-old Greg explained why his capped, empty vial floated on the water, while his uncapped vial sank. According to Greg, the cap held up the first vial! To my dismay, he verified his conclusion by removing the plastic cap and floating it on the water. How confusing! According to *my* plan, he should have noticed that the capped vial was filled with air, hence, it was lightweight; the uncapped vial filled with water, hence, it was heavy. It took a while for my supposedly flexible, adult thinking to find a way back to the objective of the experiment. "Greg, you had a good idea about the cap. It does float by itself. Let's see what happens if we put the cap on the container full of water. Now let's put the same kind of cap on the empty-looking container and watch them again." Tuning in to a child's logic makes teaching an exciting learning process for the teacher.

If children have difficulties with an activity, it may be possible that further modifications are needed. It is important not to give up on science activities because of an occasional unexpected outcome. Keep in mind Thomas Edison's observation that a mistake is not a failure if we learn how not to do it next time.

PROFESSIONAL GROWTH

Teaching can be an endlessly challenging profession when we stay open to new ideas. We can enliven our interest in teaching science by acquainting ourselves with these periodicals for children and for teachers:

- *Science & Children,* a journal of the National Science Teachers Association, 1742 Connecticut Ave. NW, Washington, DC 20009.
- *Your Big Backyard,* for preschool children; *Ranger Rick,* for primary/elementary grade children; *Ranger Rick's NatureScope,* for teachers (includes activities, extensions and outdoor projects). All publications of the National Wildlife Federation, 1412 16th St. NW, Washington, DC 20036–2266.
- *3-2-1 Contact Magazine,* E=MC Square, P.O. Box 2932, Boulder, CO 80321.

The teaching framework will achieve its greatest effectiveness when it stimulates teachers to continue growing and generating their own ideas for presenting science experiences to the children they know best.

REFERENCES

ADLER, TINA. (1991). Cumulative learning aids memory. *American Psychological Association Monitor, 22,* 10.

BRUNER, J. (1961). *The process of education.* Cambridge, MA: Harvard University Press.

CAREY, S. (1986). Cognitive science and science education. *American Psychologist, 41,* 1059–1063.

CASE, R. (1986). *Intellectual development: Birth to adulthood.* New York: Academic.

DWECK, C. S. (1986). Motivational processes affecting learning. *American Psychologist, 41,* 1040–1048.

EARTHWORKS GROUP. (1990). *50 simple things kids can do to save the earth.* Kansas City: Andrews & McMeel.

FARNHAM-DIGGORY, S. (1990). *Schooling.* Cambridge, MA: Harvard University Press.

HOLT, B. G. (1989). *Science with young children.* Washington, DC: National Association for the Education of Young Children.

JOHNSON, D. W., JOHNSON, R., HOLUBEC, E., & ROY, P. (1988). *Circles of learning: Cooperation in the classroom.* Alexandria, VA: Association for Supervision and Curriculum Development.

KYLE, W., BONNSTETTER, R., MCCLOSKEY, S. AND FULTS, B. (1985). What research says: Science through discovery—students love it. *Science and Children, 23,* 39–41.

MORGAN, M. (1984). Reward-induced decrements and increments in intrinsic motivation. *Review of Education Research, 54,* 5–30.

ROWE, M. B. (1974). Relation of wait-time and rewards to the development of language, logic, fate control: Part II, rewards. *Journal of Research in Science Teaching, 11,* (4), 291–308.

SHYMANSKY, J. (1984). BSCS programs: Just how effective were they? *American Biology Teacher, 46,* 54–57.

PART TWO

CONCEPTS, EXPERIENCES, AND INTEGRATING ACTIVITIES

4

Plant Life

They feed us, clothe us, shelter us, purify the air we breathe, and fill our visual world with beauty: the living things called plants. Children may be captivated by towering giant plants or the tiniest weed blossoms underfoot. When we share their delight, we renew our own appreciation of nature's exquisite order. The following concepts will be explored:

- There are many kinds of plants; each has its own form
- Most plants make seeds for new plants
- Seeds grow into plants with roots, stems, leaves, and flowers
- Most plants need water, light, minerals, warmth, and air
- Some plants grow from roots and stems
- Some plants do not have seeds or roots
- Many foods we eat are seeds

The first suggested experience will be limited by climate to areas where deciduous trees grow. The next group of activities calls for gathering natural materials. This may require the ingenuity of teachers in urban schools. The concluding experiences with seeds and plant growing should be possible anywhere. Suggestions for seedling care and for transplanting are included.

CONCEPT: There are many kinds of plants; each has its own form

1. Do the parts of different plants look different?

LEARNING OBJECTIVE: To observe and describe similarities and differences in the leaves, bark, and flowers of different plant sources. (Do this after a walk to collect nature materials.)

FIGURE 4–1

MATERIALS:

Large collecting bag full of found items such as:
- Leaves (2 or 3 of each kind)
- Tall grasses (include seeds or blossoms)
- Flowers
- Twigs
- Bark
- Seed pods, nuts
- Mosses, lichens

Paper lunch bags

GETTING READY:

Sort the found materials.

Fill a teacher's bag with one of each kind of item.

Distribute an assortment of materials into small bags for the children.

Place a closed bag at each place at the science table.

SMALL GROUP ACTIVITY:

1. Take one object from your bag. "Look into your bags to see if you can find a leaf that matches this one."
2. As children find similar items to compare, encourage them to notice details: "Is it just like mine? Are the tips of your leaf rounded like this one? It almost matches. Who found one with rounded tips?" (Children may have lots of information about plants already. Listen.)
3. Point out that all leaves from the same kind of tree have the same general shape. (Size and fall coloration may vary.) For example, "All sweet gum tree leaves have five points if they have finished growing. That's one way to know it is a sweet gum tree."

Note: Specimens can survive close inspection by enclosing them in a no-cost display cover. Attach well-dried material with a drop of glue to the inside of a clear plastic deli-carton or yogurt container lid. Cover with a matching lid. Join rim edges to form a case and seal with tape (Figure 4–1). (Remove dating ink by warming the lid over a cup of steaming liquid, then wiping with nail polish remover.)

2. *How do some plants rest for winter?*

LEARNING OBJECTIVE: To observe and describe seasonal changes in trees and shrubs.

MATERIALS:

Shopping bag

Old, thick catalog

Newspaper

Waxed paper

Electric iron

GETTING READY:

(Try to do this a week in advance. Save waxed leaves for another year.)

Get a good specimen leaf from each tree you plan to visit with the class.

Press leaves several days between pages of newspaper, inserted into the catalog. Weight with the iron or bricks.

Fold waxed paper over each dried leaf. Press warm iron on paper to coat leaves with wax.

Tape to a low bulletin board and label.

SMALL GROUP ACTIVITY:

1. When deciduous trees start to change color, take a tree trip. Have children circle a tree, holding hands. "What can you see above you? Below?" Repeat with an evergreen tree. (Old needles on the ground have been replaced by new ones, but not all at once.) Visit shrubs if trees are not within walking distance.
2. Supply these ideas, as needed: leaves make food for trees to grow; green stuff (chlorophyll) in leaves turns sunlight, water, minerals, and air into food; this work is finished for leaves on some trees when summer ends.
3. Gather leaves from the ground.
4. Let children put like kinds of leaves together when you get back. Ask them to try to match their finds with the mounted specimens. Encourage them to bring leaves from home to try to match them.
5. Save the surplus leaves for art activities or for compost.

Group Discussion: Recall the fun of collecting leaves. Ask what happens to the leaves from deciduous trees and shrubs that rest for winter. Introduce the idea that when chlorophyll goes out of these leaves, other colors that are also in the leaves show instead of only the green chlorophyll. (You might be able to find mottled leaves that fell before all the chlorophyll left.) Talk about how the leaves can still be valuable after they fall. If possible, start a compost bag or heap (see pages 67–68).

CONCEPT: Most plants make seeds for new plants

1. *Are there seeds in fruit?*

LEARNING OBJECTIVE: To become aware that most plants form seeds which may grow into new plants of the same type.

FIGURE 4–2

MATERIALS:

Any available seed pods: flower, tree, shrub, tall grass seed head

As many of these as can be brought in: apple, tomato, pomegranate, peach, ear of corn in husk, orange, nuts in shells, melon, squash, green beans, apricot, cucumber

Clean meat trays

Paring knife (*for adult use*)

Smocks

Newspapers

Nutcracker, if needed

GETTING READY:

Wash fruits and vegetables.

Cover table with papers.

Hand washing for all participants.

Put out plants with seed pods first to minimize confusion.

SMALL GROUP ACTIVITY:

1. Show a seed pod. Open it. (Pods still attached to stalks are best.) Let children tell what they know about it. "Each kind of plant has a job to do: to make seeds for new plants just like it. Each plant forms seeds when its flowers stop blooming. Some seeds are protected by covers we like to eat. Let's try to find seeds inside these fruit and vegetable covers."
2. Take plenty of time to decide if seeds will be inside each fruit. Look at stems and blossom ends. Cut open and share tastes and sniffs with everyone. (Figure 4–2).
3. Save melon and squash seeds. Later let children wash them and dry them on trays. Save for bird feeding trays. If corn is fresh, pull back husks and hang to dry. Let children shell dried corn for the birds and for a spring sprouting project. Tack one dried husk and ear of corn to a tree for the birds.

Read *The Apricot ABC,* by Miska Miles. This is a lovely poem about the cycle from fruit, to tree sprout, to fruit. Try to read it without referring to the alphabet letters hidden in the charming illustrations. The story line may fade in importance if children concentrate on hunting for the letters.

Carrie examines the shower of seeds that spilled from the iris pod.

2. How are seeds scattered?

LEARNING OBJECTIVE: To examine how different kinds of seeds are formed and scattered.

MATERIALS:

Locally available seeds from: garden plants; weeds (teasel, milkweed, burdock); grasses (wheat, oats); and trees (ailanthus, locust, oak, pine*, chestnut)

Magnifying glasses

Old fuzzy mittens and socks

Trays

SMALL GROUP ACTIVITY:

1. Arrange materials on trays. Encourage children to shake seeds from pods, brush hairy seeds against the mitten, beat grass seed spikes against trays to release grains, and take a close look at burrs and hairy seeds with the magnifying glass to see tiny hooks on the tips.
2. Take an envelope of winged seeds and a few heavier nuts to the playground. Let children launch them from a high place. Compare what happens to each kind of seed.

*Seeds lie beneath separate cone scales. Old cones on trees may no longer contain seeds. Tightly closed new cones will dry and open in a warm oven with the heat turned off. Seeds can then be found.

GETTING READY:

Gather ripe seed stalks in advance. Store in open containers or hang in tied bunches to dry. Preserve husks and pods intact. Find weeds in vacant lots or in roadside ditches.

Group Discussion: Talk about how different seeds came out of their coverings in different ways. "Have you ever seen plants growing in places where people couldn't

Tiny hooks in the teasel seedhead cling to Lori's mitten: a "handy" way to travel.

plant them—in sidewalk cracks or rock crevices? How could seeds get there? How many ways can we think of?"

Listen to Tom Glazer, "How Do the Seeds of Plants Travel?" on *Now We Know (Songs to Learn By)*.

CONCEPT: Seeds grow into plants with roots, stems, leaves, and flowers

1. What is the secret in a seed?

LEARNING OBJECTIVE: To examine and notice the parts of seeds.

MATERIALS:

Dried lentils, lima or navy beans

Peanuts in shells

Desirable if available: maple tree seeds, avocado, fresh green beans or peas

Magnifying glass

GETTING READY:

Soak seeds overnight in enough water to cover. Keep a few seeds dry for comparison.

LARGE GROUP ACTIVITY:

1. Comment, "There's a secret in every seed. Let's find it." Carefully slip off a seed coat. Pull apart the 2 parts (cotyledons) that supply food for starting a new plant. Find the "secret": the tiny new plant ready to start growing (embryo).
2. Let the children continue to open seeds and find new plants. Offer the magnifying glass for closer inspection.
3. If a very ripe avocado is available, slice the fleshy part in half. Twist slightly to pull apart. Examine seed with children. Peel the seed coat at the base. A ripe fruit seed may already be split, revealing a root tip. Do *not* split it open.

To Start an Avocado Tree: Slice away about ¼″ (1 cm) from the base of the avocado seed. Insert three round toothpicks midway through the seed. Suspend the seed in a jar of water. Keep in a warm place away from direct sunlight. Change the water weekly. After the root appears, move the jar to a sunny location. When leaves appear on the stem, gently plant in a pot at a depth of approximately 5″ (13 cm). Leave the top quarter of the seed exposed above the soil. Water at least every other day. Spray or wash leaves frequently.

2. *How do seeds start to grow?*

LEARNING OBJECTIVE: To observe initial seed sprouting.

MATERIALS:

Method 1

Matching disposable plastic tumblers

Transparent tape

Cotton balls

Dried legumes: navy or lima beans, lentils (fresh stock)*

Water

Plastic prescription vial

Desirable: mung beans (natural food store)

Method 2

Plastic sandwich bags

White paper toweling

Stapler

Masking tape

SMALL GROUP ACTIVITY:

Method 1

1. Make a sprouting dome: Wet 4 or 5 cotton balls, press out excess water, and flatten. Line the bottom and sides of one tumbler with cotton.
2. Let children see and feel the beans. Recall the secret inside the seeds as they help place 4 beans between the cotton and the tumbler side. (Try to use more than one kind of legume to see which sprouts first, which gets tallest, which grows for the longest time.)
3. Upend the matching tumbler on the rim of the prepared tumbler. Tape the rims to make a dome enclosure (Figure 4–3).
4. Place it away from direct sunlight where temperature will be even.
5. Put one of each kind of seed in the vial for later comparison. Start a calendar record of starting date, first root, first stem, and first leaf appearance. Crayon-mark daily growth level on the tumbler.

Method 2

1. Each child makes a sprouting bag. Fold toweling to fit bag and dampen. Place 5 lentils on damp toweling; staple bag shut. Label with child's name on a piece of tape.
2. Children keep daily sprout growth logs by sketching or writing descriptions, and by measuring root and stem length.

*Two common causes of germination failure are old seeds and an overheated, dry room. Don't expect goods results with seeds of unknown vintage, nor with uncovered sprouting containers.

Note: It's a good idea to start two germinating domes. Keep one available for children to pick up for a close look. If sprouts don't survive the inspection, the other dome will be available. It's hard to only look when leaves are showing beneath a rakish seed cover cap.

FIGURE 4–3

3. Which direction do roots and stems grow?

LEARNING OBJECTIVE: To observe the tendency of roots to grow downward toward water and of stems to grow upward toward light.

MATERIALS:

Same as for the seed sprouting experience

SMALL GROUP ACTIVITY:

1. "Notice which direction the seedling roots and stems take. Is it the same for each seedling?"
2. Gently turn one seedling so that the stem points down and the root reaches up. Mark an *X* on the glass beneath it.
3. Check each day for changes in root and stem growth direction. Look at the cotton behind the seeds. Roots may poke down into it toward water.

Group Discussion: Ask children if trees grow with their branches and leaves in the soil and their roots in the air; if flowers blossom underground, or if plants send roots into the ground and other parts into the light? Why is this so? Help the children recall that leaves need light and air to perform their food-making job. Roots have the job of getting water and minerals from the soil so that the plant can live and grow. The upended seedling root and stem twisted and turned to grow in the directions where each could get what it needed.

CONCEPT: Most plants need water, light, minerals, warmth, and air

1. Can we raise plants from seeds?

LEARNING OBJECTIVE: To become aware of and to provide the elements needed to promote seed germination and plant growth.

MATERIALS:

Zinnia or marigold seeds, package-dated for current year

Small package commercial potting soil (sterilized)

Teaspoons

Eggshells (as large as possible), one for each child, plus spares

Frozen pie pans with plastic dome cover intact*

Water

Sand or fine gravel

Punch-type can opener

Toothpicks

Medicine droppers

GETTING READY:

Collect eggshells in advance.

Cover table with newspapers.

Put an inch of sand in each pie pan.

Make tiny paper name tags and staple to, or lace with, a toothpick.

SMALL GROUP ACTIVITY:

Let the children:

1. Fill eggshells almost to the top with soil (cup a shell in one hand while filling).
2. Push shell gently into the sand-filled pie pan. Pans can hold about 12 shells.
3. Use dropper to dampen soil well. Place one seed on soil, then cover with spoonful of soil, press firmly, water again, and insert name tags in shells. Plant extra shells to replace possible failures.
4. After all shells are in place, tape plastic domes to pie pans. Place in a spot away from drafts, radiators, and direct sunlight.
5. Start a calendar record of seedling growth stages.
6. Set aside a seedling dampening time every other day. Ask, "If you were a tiny new plant, how would you want to be cared for?"

*If frozen pie containers are not available, use two egg cartons, one inside the other. Slide cartons into plastic bags and fasten ends with rubber bands.

Seedling Care: After leaves appear, provide moderate light, such as a northern exposure. Cover only at night to retain moisture. Allow children to water their own plants with the dropper. It's hard to overwater this way, but if it happens, blot up standing water with absorbent materials. Check each shell before covering pans at night in case someone forgets to water a plant.

Transplanting: For several days after the second pair of leaves appear, give seedlings a few hours of direct sunlight, preferably outdoors in a sheltered spot. Pack the shells in small milk cartons stuffed with crumpled paper for children to carry home. Tell parents that the eggshells can be planted directly into the ground if the soil is warm. Crushing the shells with thumb and fingers as they are placed helps roots become established.

If many children in the class are apartment dwellers, you may want to choose dwarf plant varieties to grow. Transplant the seedlings before sending them home. Use any container large enough to hold about three cups of soil. Punch a drainage hole in the bottom of cartons or cans, add a layer of small rocks, and fill ⅓ full with soil. Crush the shell, plant firmly, cover with more soil, and water it.

About Soil: To control one possible source of failure, commercially prepared potting soil is recommended for germinating seeds. For other classroom use, let children enjoy mixing their own potting soil: ⅓ ordinary soil, ⅓ sand, ⅓ peat moss. Talk about how soil is made of crumbled rocks, dead plants, and insect matter.

Group Discussion: Compare the calendar records of the germination dome and the plant-raising experiences. "Were seeds treated alike in both experiences? Which seedlings stopped growing and withered? Which ones kept growing?" Moist seeds can grow only until their built-in food supply is used up. Plants rooted in soil can use minerals and moisture from the soil to help the leaves make their food for growth. Would the seeds grow in a freezer? Find out.

Additional Experiences: Keep some soil on hand all year to be ready for planting opportunities. Plant seeds used in the earliest investigations of seeds. Fresh peas or beans can be planted directly into the soil. Corn must be dry enough to be pried off the cob without breaking open. Plant some jack-o-lantern seeds after Halloween. Try growing plants from grapefruit seeds. Late season, tree-ripened grapefruit seeds seem to respond most promptly. Soak them in a small amount of water until the seeds sink to the bottom of the container (about three days) before planting. Try fresh date pits. A general rule for planting is that seeds should be planted at a depth twice the width of the seeds.

2. *How do plants take up water?*

LEARNING OBJECTIVE: To observe how moisture is taken into the stem of a plant.

MATERIALS:

2 stalks of celery, with leaves

2 jars

Food coloring, blue or red (enough to make dark color)

Water

SMALL GROUP ACTIVITY:

1. "How does water get into the leaves of a plant?"
2. "Let's see if we can figure out how water moves up these stalks of celery." Let children stir food coloring into one jar of water.
3. Check within an hour for signs of color in leaf tips. Separate a dyed tube from the stalk so the dye can be seen in whole length. Slice a cross section from the bottom of the stalk to examine.
4. "What do you think would happen to a stalk of celery if it had no water for a while? Let's find out." Leave the other stalk in the empty jar overnight. Check its condition the next day.
5. "Do you think water will change this stalk? Let's try it." Add water to the jar. Check it the next day. Has it revived? Clarify that celery plants have roots in the ground when they are growing. Roots take up water from the soil and the water travels up through the stalk tubes.

CONCEPT: Some plants grow from roots and stems

1. *Can plants grow from potatoes? carrots?*

LEARNING OBJECTIVE: To observe a different means of developing a new plant: by sprouting roots or stems.

FIGURE 4–4

MATERIALS:

White potato, or a sweet potato (do not try a kiln-dried potato, local ones may be untreated)

Onion

Carrot (fresh, an aged one with root hairs showing won't do much)

Ketchup bottle cap

Disposable plastic tumbler

Saucer

Glass jars

Water

Round toothpicks

SMALL GROUP ACTIVITY:

1. "Some plants can be grown from parts of the stem or root. A potato plant will start this way. Let's see if we can start one growing in water."
2. Stand, or suspend from toothpicks, a potato in a jar of water. (About ⅓ of the tapered end of a sweet potato needs to be in water.) Add to or change water as needed during the growing period. What happens?
3. "What do you think feeds the vine?" (Roots and tubers provide stored food for new growth.)
4. Keep a calendar record of growth. A successful sweet potato vine will flourish for months before its food supply is depleted. It may be planted in the ground when the old potato starts to cave in.
5. Cut the carrot top twice the bottle cap depth. "Perhaps this carrot top could feed a small new plant. Let's keep it wet and see what happens." Put it in the bottle cap; add water. Place on a saucer and invert the tumbler over the carrot top (Figure 4–4). Place in a light area, avoiding direct sunlight. Add needed water as long as the ferny plant continues to grow.

Group Activity: Try to plant some daffodil bulbs outdoors in fall. Later, if you should find a sprouting onion in your kitchen, slice it open vertically so the children can see the new plant tucked inside its food supply bulb. Talk about the bulbs outdoors waiting for the spring sunshine to warm the earth and the rain to start them growing. Try to start a few paper narcissus bulbs in the room. Follow package directions for growing conditions.

CONCEPT: Some plants do not have seeds or roots

1. What is a mold?

LEARNING OBJECTIVE: To observe the ways simple plant forms grow and develop.

Group Discussion: Bring a piece of bread, two screw-top jars, and a drop-top bottle (like a soy sauce bottle) of water to a group gathering. Recall with the children that most plants make new plants from seeds or root and stem parts. Say that a few kinds of plants grow from tiny dust-speck bits called *spores.* Millions of spores blow about in the air, but they are too small to see. When spores land on a warm, moist food source, they grow into plants that we can see.

"Perhaps we can grow a plant like that on a bit of bread." Put half of the bread in each jar. Cover one jar. Let the children sprinkle water on the other piece of bread. Leave this jar uncovered for an hour. Then cover it and store it in a warm, dark place for a few days. Compare the two pieces of bread. Do they look the same? Leave the moldy bread in the jar to develop a luxuriant fur coat (perhaps the black spore clumps will be visible as specks on the mold).

If you want to develop this concept further, experiment with the growth requirements of yeast. Examine mushroom gills, watch for algae and lichen growing on rocks, sample Roquefort or bleu cheese, watch for brown or green algae growth in an aquarium, bring in mosses and ferns to plant in a terrarium. The green spore-producing plants can make their own food. The colorless spore plants must use dead plants or other sources for food.

Read *Lots of Rot,* by Vicki Cobb, to teach children how molds help reuse the materials in the ecosystem.

More experiments with molds and decomposers can be found in *Biology for Every Kid,* by Janice Van Cleave.

CONCEPT: Many foods we eat are seeds

1. Which seeds are good to eat?

LEARNING OBJECTIVE: To experience firsthand that many seeds are used for food.

MATERIALS:

For grinding:

Blender

Cracked wheat

Peanuts

Peppercorns

Pepper mill

For Cracking:

Peanuts in shells

Whole sunflower seeds

For Toasting:

Hulled sunflower seeds

Hulled pumpkin seeds

Squash seeds

Electric skillet

SMALL GROUP ACTIVITY:

Grinding: (To show how grain becomes flour.) Follow blender instructions to grind wheat. Use blender book recipe to make peanut butter.
Toasting: Toast hulled sunflower and pumpkin seeds in an electric skillet at low heat with a few teaspoons of margarine. Stir until lightly browned. Sprinkle with salt. Delicious and nutritious.
Cooking: Follow package directions. Compare dry and cooked foods, then eat.
Popping: Mention the bit of moisture in dry kernels that changes to steam, exploding the kernels when heated. Plan carefully to ensure safe use of electrical equipment in the classroom.
Sprouting: See page 53.

For Cooking:

Rice

Oatmeal

For Popping:

Popcorn, salt, margarine

Transparent dome popper

For Smelling:

Whole seed spices like nutmeg, corriander, anise, poppyseed, or peppercorns

Read *The Popcorn Book* by Tommie dePaola.

Group Activity: Open a coconut, one of the biggest seeds that grows. Puncture soft spots at the end of it and drain the juice into a measuring cup; then let children sample spoonfuls of juice. Crack the shell with a hammer and scrape out small pieces of coconut for children to taste. If a blender is available, grate some of the coconut according to manufacturer's instructions to make this confection:

Seed Candy

Use the blender to grate: 1 cup of coconut
1 cup hulled sunflower seeds

Mix with: 2 tablespoons peanut butter
2 tablespoons honey, confectioners sugar, or maple syrup (both of the latter come from plants)

Form into a log, slice thinly, and enjoy.

Go on with other foods from plants as long as your enthusiasm lasts. Try, or talk about, root, stem, blossom, or leaf parts of plants that we eat.

Safety Note: Do not suggest that children pick wild mushrooms. Do not give these things to children: castor beans, berries from pokeweed, nightshade, bittersweet, yew, holly, privet, or mistletoe. Do not allow experimental eating of buckeyes, daffodil bulbs, iris or lily tubers, or poinsettia leaves. Further information about poisonous plants can be found in *A Guide to the Medical Plants of the United States*, by A. and C. Krockmal.

PRIMARY INTEGRATING ACTIVITIES

Music (Resources in Appendix 1)

Try these children's songs about plants.

1. Sing Raffi's "Popcorn".
2. Sing "The Tree in the Woods," the old English cumulative song about a tree, its parts, and its occupants.
3. Vary "Oats, Peas, Beans, and Barley Grow" to sing about vegetables that might be more familiar to some children: "Corn, peas, beans, and lettuce grow . . ."

Natural science and natural math coincide when the group counts and sorts field trip finds.

Plants as Music Makers. Dried gourd maracas are well-known plant instruments. Drums or resonant rhythm instruments can be made from hollow logs and coconut shells. Don't forget willow whistles, dandelion or grass stem whistles, wooden flutes, and recorders. Try to show a set of wooden tone blocks.

Math Experiences

Keeping Records. Take a shopping bag on nature walks to hold things found along the way. Use these things in the classroom for sorting, classifying, and counting experiences. Record the results on a bulletin board or an experience chart (*We Found:* 12 leaves, 10 acorns, 8 bits of bark, 1 piece of moss). For younger children, use numerals and samples of the material; for beginning readers, use numerals and words.

Accumulate Collections. Chestnuts, acorns, pinecones, sweetgum balls, and buttonwood "buttons," and sticks can be collected. Use them as materials to sort, match, count, and weigh. Use them as markers in bingo games. Group them on meat trays to make sets to match numeral cutouts or numerals marked on cardboard squares. Arrange sticks in order according to their length.

Commercial materials for number experiences include magnet-board apples, and felt tree and fruit cut-outs for flannel boards. Puzzles to arrange in time sequence include the subjects of seed germination and maple sugar making.

Recording Growth. Use string to record plant growth. Hold the string beside the plant stem. Cut string at the level of the stem height. Tape the strings each day to a poster board chart, and label with the date to keep a visual record of the plant's growth.

Classifying Seeds. Give children a jar of varied dried beans and legumes to sort into ice cube trays. Have children describe differences and similarities of the seeds.

Stories and Resources for Children

BRANDT, KEITH. *Discovering Trees.* Mahwah, NJ: Troll, 1982. From seed to tree, plus seasonal aspects of the growth cycle, are narrated at preprimary child's interest level. Paperback.

BRANLEY, FRANKLYN. *The Sun: Our Nearest Star.* New York: Harper & Row, 1988. Includes experiments demonstrating that energy from the sun helps plants grow. Paperback.

BROWN, MARC. *Your First Garden Book.* Boston: Little, Brown, 1981. A delightful book for young city gardeners. A sidewalk crack is planted with fork and shoehorn for tools.

BURNIE, DAVID. *Plant.* New York: Alfred A. Knopf, 1989. Beautiful color photograph layouts and clear information in this London Museum of Natural History Eyewitness Series. Intended for older audiences, but fascinating for primary-grade children as well.

______. *Tree.* New York: Alfred A. Knopf, 1989. Eyewitness Series.

COBB, VICKI. *Lots of Rot.* J. B. Lippincott, 1981. The mildew plant is shown as a nuisance *and* a necessity. Amusingly illustrated.

DAMON, LAURA. *Wonders of Plants and Flowers.* Mahwah, NJ: Troll, 1990. A broad look at plant life, from the microscopic to the gigantic. Upper primary grade children. Paperback.

DEPAOLA, TOMIE. *The Popcorn Book.* New York: Holiday House, 1978. Two children acquire interesting lore about this favorite food, as they prepare a huge batch for themselves. Paperback.

HELLER, RUTH. *The Reason For A Flower.* New York: Grosset & Dunlap, 1983. Strikingly beautiful, fascinating illustrations by the author, and a simple text make this a book that children will return to and remember.

KRASILOVSKY, PHYLLIS. *The Shy Little Girl.* Boston: Houghton Mifflin, 1970. A shy girl finds a best friend through shared fun with dandelion chains.

KRAUSS, RUTH. *The Carrot Seed.* New York: Scholastic Book Services, 1971. A simply story of a child's faith in a seed.

LOBEL, ARNOLD. *Frog and Toad Together.* New York: Harper & Row, 1972. Toad has some funny ideas about coaxing seeds to grow.

______. *The Rose in My Garden.* New York: Greenwillow, 1984. A cumulative tale in lovely prose that takes place among garden flowers. Charming illustrations by Anita Lobel.

MARCUS, ELIZABETH. *Amazing World of Plants.* Mahwah, NJ: Troll, 1982. Concise, general botanical information makes this a good reference for primary grades. Paperback.*

MCMILLAN, BRUCE. *Apples, How They Grow.* Boston: Houghton Mifflin, 1979. A series of black-and-white photographs trace the development of the apple from winter bud to harvest. Large photographs accompany simple text for children; detailed photographs and detailed information for the adult reader.

MILES, MISKA. *The Apricot ABC.* Boston: Little, Brown, 1969. Beautifully illustrated story of insects' responses to a fallen apricot that eventually becomes a bearing tree.

MINER, IRENE. *Plants We Know.* Chicago: Children's Press, 1981. Well-illustrated, organized information about plants for young readers.*

NEWTON, JAMES. *Forest Log.* New York: Thomas Y. Crowell, 1980. A beautiful introduction to the biological community concept for the primary grades.

ROCKWELL, ANNE & HARLOW. *How My Garden Grew.* New York: Macmillan, 1982. A charming record of a young gardener's year. For preschool children.

*Starred references, written at the young child's level of understanding, can help teachers with minimal backgrounds in science expand their knowledge base.

ROMANOVA, NATALIA. *Once There Was A Tree.* New York: Dial, 1989. This lovely ecosystem tale from Russia is enriched by soft, detailed Dürer-like illustrations. New York Times Notable Books of the Year award. Paperback.

SCHWARTZ, JULIUS. *Magnify and Find Out Why.* New York: Scholastic Book Services, 1972. A close-up look at roots, seeds, and the seeds we use for spices.

SELSAM, MILICENT E. *Seeds and More Seeds.* New York: Harper & Brothers, 1959. Benny learns firsthand about seeds, beginning with his inquiry about whether a seed, a stone, or a marble will grow. He learns how seeds are formed and how they scatter.

SILVERSTEIN, SHEL. *The Giving Tree.* New York: Harper & Row, 1964. A boy grows up with a tree that gives him shade, apples, shelter, and finally a resting stump.

VAN CLEAVE, JANICE. *Biology for Every Kid.* New York: John Wiley, 1990. Includes simple experiments with bean seedlings and molds.

WYLER, ROSE. *Science Fun with Peanuts and Popcorn.* New York: Simon & Schuster, 1989. Seeds to grow and seeds to eat! This book combines science with the child appeal of favorite foods. Paperback.

ZION, GENE. *Harry By The Sea.* New York: Harper & Row, 1965. A funny story about seaweed that turns Harry into a temporary sea monster.

Poems (Resources in Appendix 1)

The collection, *Poems to Grow On,* compiled by Jean McKee Thompson, has four poems about seeds appropriate to read during the seed investigations:

"The Little Plant," by Kate Louise Brown
"The Seed" and "Carrot Seeds," by Aileen Fisher
"Seeds," by Walter de la Mare

Fingerplays

This traditional finger play fits well with the concept that living things reproduce in their own special form.

The Apple Tree

Way up high in the apple tree	(Stretch arms up)
Two red apples, I did see.	(Make circles with hands)
I shook that tree as *hard as I could.*	(Shake "trunk")
Mmmmm, those apples tasted good!	(Pat tummy)

—AUTHOR UNKNOWN

Substitute "orange carrots, green pears, two bananas," and so on, for "two red apples, I did see." The children will enjoy catching and correcting your mistake. Ask them why it must be apples growing on the apple tree. "Really? Don't carrots grow on apple trees? Then, where do they grow? Do they grow in the ground from apple seeds?"

My Garden

This is my garden.	(Hold one hand palm upward)
I'll rake it with care.	("Rake" with curled fingers of other hand)
Here are the seeds	
I'll plant in there.	(Pantomime planting, seed by seed)
The sun will shine.	(Circle above head with arms)
The rain will fall.	(Fingers flutter down)
The seeds will sprout	(Spread fingers of one hand. Push up other fingers between them)
And grow up tall.	(Bring hands and forearms together. Move up, spreading palms outward as arms move up)

—AUTHOR UNKNOWN

Art Activities

Collage. Dried grasses, leaves, pressed flowers, flat seeds, and small twigs make lovely collage materials. Tape may be needed when younger children include twigs and long grasses in their work. Vary the background colors to bring out the hues of the dried materials. Flat macaroni wheels, spaghetti, lentils, and similar foods can be used in plant life collages, after investigating plants that we eat.

Printing. Carve simple shapes from pieces of carrot or potato, or use halves of green pepper or citrus fruit as printing blocks. Make a simple stamping pad for the activity by folding paper towels into a plastic meat tray and pouring liquid tempera on them.

Rubbings. Tape a single fresh leaf or a pattern of small leaves to the table in front of each child. Cover with a sheet of heavy paper. Let children rub a crayon over the paper to bring out the relief design of the leaf veins.

Translucents. Fresh leaves, flower petals, and grasses can be sealed between two sheets of waxed paper to make translucent window hangings. Do not use an electric iron in the classroom to seal the paper. For safety, and for maximum child participation, use a newspaper-covered electric food warming tray as a heat source. Give children a pizza roller or a child-size rolling pin to apply light strokes of pressure to the waxed paper.

Easel Painting. Cut newsprint sheets into very large blossom shapes or leaf shapes for the children to paint at the easel. A border of painted flowers makes a charming room decoration for a special occasion.

Cut-Paper Mural. Offer strips of green paper for stems and leaves and assorted sizes of circles in blossom colors. Indicate ways to change the texture and shape of the rounds such as notching; slashing for fringes; scoring with a ruler edge and pinching into cup shapes; cutting into spirals; and curling edges around a pencil. Put out scissors and white glue or paste. Assemble the completed fantasy flowers into a beautiful mural for your classroom door.

Leaf Mobile. Let children cut out freehand leaf shapes (or whatever satisfies them as appearing leaflike). In spring, use green paper; in fall, use orange, brown, and red. Use 5-inch squares of paper. Suggest making the job easier by folding the paper in half. Let them punch a hole in each leaf and thread it with a small bit of yarn. Help them tie or tape their leaves to an interestingly shaped branch. Hang the branch from a ceiling beam, or staple it to a bulletin board with some of the twigs extending into the room.

For more nature art ideas consult *Puddles, Wings and Grapevine Swings,* by Imogene Forte and Marjorie Frank.

Dramatic Play

Farming. Playgrounds that offer a bit of shady ground for digging are natural settings for spontaneous farm play. Provide children with small, sturdy rakes, hoes, shovels, buckets, and a wheelbarrow. They will find rocks or cones to plant, and leaves, grasses, or pine needles to harvest without further suggestion. Children will need to know the boundaries of the permissible digging area. Listen to their planting ideas.

Sand Table Indoor Gardens. A collection of twigs, pinecones of several sizes, dried grasses and pods, or perhaps wildflowers, can be arranged by the children to create small landscapes in the sand table. (Dampened sand will hold better than dry sand.) Rubber toy animals and people can be added.

Blossom Fun. If you can find an abundance of wildflowers for children to gather, use the blossoms to make beautiful (though short-lived) decorations. Show children how to make a split midway in a dandelion stem in which to insert another dandelion stem, and so on. The resulting rope can be looped into necklaces or crowns, or be allowed to get as long as possible. Leaves and sturdy blossoms can be threaded on soft, covered wires from a telephone cable to make bracelets. Large leaves strung together can become Indian headdresses. Two hollyhock blossoms can be stacked and joined through the center with a toothpick to make dolls with hollyhock-bud heads and daisy hats.

Children can place small sprigs of evergreen in a water-filled ring mold to make a tabletop Christmas wreath. Use small pinecones for turkey bodies; tiny hemlock cones, painted red, for heads; bird feathers for tails.

Food Experiences

Make applesauce when apples are in season. Try drying raisins if grapes are abundant in your area. Children can learn to scrub and peel raw fruit or vegetables for snacks and prepare vegetables for soup. Any time flour or sugar is used in classroom cooking projects, mention that plant source.

Discuss the parts of plants that are being served for lunch. "Are we eating the leaf, the root, or the stem of the celery plant (the potato, the carrot)?" Talk about the wheat seeds and the stems of sugar cane plants or the roots of sugar beet plants, that went into the snack-time graham crackers.

Thinking Games: Logical Thinking and Imagining

What If? What if you were as tiny as your thumb? Which plants would you want to live near? to sleep in? to use for food? to hide under when it rains? to climb for fun? Read "The Little Land," by Robert Louis Stevenson, for inspiration (Appendix 1).

Play a simple guessing game in which children take turns describing mounted pictures of plants for others to guess. "I'm thinking of a plant that we eat for breakfast, makes something for blue jays to eat on the feeding tray," and so on. Seed catalogs are good sources for pictures. (Pictures printed directly on cardboard food boxes do not need further mounting.) Include a piece of plain paper, a mounted ice cream stick, and a cotton ball or fabric.

Field Trips

We usually think of pleasant woods and meadows as ideal sites for plant life field trips. But the closest grocery store also has important plant learnings for children. Go there to see how much we depend on plants for food. Visit the produce section; the shelves of dry staple foods (flour, pasta, sugar, legumes, cereals, and all the packaged mixes); the canned and frozen fruits and vegetables; and the baked goods. A natural foods shop (not a health food store that features vitamin preparations) is a fine place to see what whole grains look like, and to watch a small flour mill in operation. Flower shops and greenhouses can be fascinating, but make sure children are welcomed by the owners.

Indoor/Outdoor Tree Walks. How many different kinds of trees grow within walking distance of your school? How can children tell that they are different? If leaves are too high to be examined closely before they fall, bark characteristics can provide identification clues. Let children make bark rubbings with old crayons on thick, flexible paper. The backs of vinyl-coated wallpaper samples are good for this. Compare the differences in bark textures for different species.

Now move indoors to complete your tree walk. Ask the children to tour the classroom looking for all the wood they can find in use there. Old buildings may have ceiling moldings and window and door frames of wood, so suggest looking up to find wood. Make a list of the found items. Pencils and paper should be on the list, too.

Having a Quiet Look Outdoors. When the weather is good, explore a nearby weedy patch. Minimize the scatter tendency of unconfined children by defining small observation spaces with 6-foot loops of yarn. Before going outdoors decide on groups of three or four children to share a looking loop. Children can stretch out on the ground, radiating from the loop like the spokes of the wheel. Ask children to report what they see in their space. Later make a sum-

Hollis checks pussywillow developments for his "Signs of Spring" story.

mary chart of the observations, including soil, rocks, and small creatures, as well as plants. Tape samples to the chart.

Wildflower Gathering. Check grassy areas near your school in early spring for the presence of tiny wildflowers. Try to obtain permission for your children to pick tiny bouquets to take home. (Small groups make it a happier occasion.) Show the children where they may hunt, then let them take their time to find spring beauties, chickweed, violets, or whatever the lawnmower spares. Put the flowers in capsule vials of water. Label each with the child's name. Use them for table decorations, then wrap each child's flowers in a twist of waxed paper when it is time to go home.

Creative Movement

Curl up on the floor with the children, pretending to be seeds that have been planted in spring (or bulbs planted in fall if this was a class project). Move with the children to enact the growing story as you softly tell it. "Here we are waiting under the ground. The sunshine makes the soil warm; rain falls, and we begin to expand. The tiny plant inside grows bigger and pokes out of our seed cover. We send a root down to get water. Now our stem starts to push its way up . . . up . . . up to find the sunlight." Slowly describe the growth of the plant above the ground: leafing; budding; blooming; swaying in the breeze; feeling the sun and rain; losing petals; forming seeds; then slowly withering and scattering seeds for next year's plants.

SECONDARY INTEGRATION

Keeping Concepts Alive

A row of potted plants on the classroom windowsills guarantees year-round attention to plant growth needs. Plant tending can occupy an honored position on the children's daily job chart. Younger children may need some help carrying the pitcher and deciding how much water to use. Older children can handle the task independently if pots are labeled with suitable watering advice. Comment on changes taking place, such as new buds, fading leaves, and unwanted insect tenants.

Classes in session during the regional growing season might be able to keep plant life concepts in focus by planting and maintaining a garden. Other classes could try portable gardens: plants started at school in the spring, moved to the home of a child or teacher for summer care, and returned to school for a fall harvest—with luck. Start pumpkin and sunflower seeds in good soil, using 2-gallon plastic buckets with drainage holes cut in the bottom. Thin out all but one vigorous seedling. Give it full sunshine and plenty of water. The plants won't attain full growth, but the pumpkin can produce a vine with leaves, tendrils, blossoms, and possibly a fledgling pumpkin. The sunflower will develop a seed head and may grow taller than a 5- or 6-year old. Slip a section of old hosiery or netting over the fading blossom to keep the birds from feasting on the seeds too soon.

If long-term growing projects cannot be worked out, perhaps you can duplicate the efforts of one very fine teacher of 3-year-olds. Each fall she scouts the countryside to find a whole, dried cornstalk complete with ears of corn. It creates a fall harvest mood in her room, and provides an awesome lesson for the children who seem dwarfed by the giant that grew from a single kernel of corn. Later, shelling the corn becomes an absorbing task for the children. They feed some kernels to the birds, and save some to germinate in spring. Other plant life experiences sometimes come about inadvertently. Capitalize on them. For example, an aging jack-o-lantern might develop a moldy spot. Instead of quietly discarding it, keep it in the room (as long as odor permits) to observe the changes. Then add it to a compost pile so that it can contribute soil enrichment for next year's plants. Read *Lots of Rot* by Vicki Cobb.

Relating New Concepts to Existing Concepts

Soil Composition Relationships. The natural cycle of renewing limited materials in the ecosystem is well illustrated in the creation of fertile soil. Pulverize shale to form powdery clay (see chapter 10). Try growing an extra seedling in it. Compare its growth with seedlings growing in true soil that also contains bits of decayed plant and animal matter. This idea could lead to starting a compost pile or bag in the fall. Let the children rake available leaves and grass clippings into a large plastic trash bag. When it is approximately half full, add a soup can full of fertilizer or powered lime, a can of water, and a few shovelfuls of dirt. Close the bag and leave it outdoors where it won't be forgotten during the winter. It will need to be turned, shaken, and opened several times for air. Although the completeness of the change that may occur by spring is not predictable, bacterial growth within the bag environment will have interacted with the materials to promote decay into compost (humus) to enrich the soil.

If molds appear on some of the decaying materials, examine them with magnifiers. Talk about molds as tiny, leafless plants that use other materials as food. Vicki Cobb's book, *Lots of Rot*, describes the value of molds as decomposers.

Discuss how completely rotted natural materials improve the soil and make it possible for strong, new plants to grow and produce food. Talk about renewing the soil this way as one of the wonderful cycles of nature: from living plants to decomposition . . . from enriched soil to healthy living plants again . . . and again . . . and again. It is comforting to understand the concept that once-living things continue to have a function after they die. (Furman. See resources)

Plants That Help Crumble Rocks. Look in shady areas for greenish-gray patches of lichen growing on rocks. Lichen are fungi and algae that live together as a single unit. Together they make acids that slowly dissolve the rock surface.

Other plants sometimes grow in rock crevices, and their roots may break off pieces of rock. Perhaps there is a section of concrete sidewalk near your school that has been cracked or pushed up by strong tree roots. Watch for them as you take walks with children. Stop to look at them. Recall with the children that slowly crumbling rocks become part of the soil that plants grow in (see resources, Chapter 10).

Air/Water Cycle. Use the term *evaporation* when plants are being watered. Some of the water will be taken up by the roots of the plants; the rest will evaporate (see resources, Chapter 8).

If you make a terrarium, talk about how rarely it will need to be watered. Bring out the idea that the water taken up by the air in the closed terrarium will change into large drops on the cool glass sides. If you make the terrarium before the children have tried the evaporation and condensation experiences, postpone the discussion.

To make a small terrarium from a 2-liter plastic soda bottle, begin by pouring hot water into the bottle to soften the glue adhering the base and label to the bottle. Use a table knife to separate the loosened base from the bottle. Cut off the neck portion of the bottle so it will fit, inverted, into the base. (Do these steps away from children.) Put a layer of gravel and some charcoal bits in the bottom of the base. Add a layer of potting soil. Arrange small plants in the soil. Dampen the soil; do not saturate. Fit the inverted dome firmly into the lid to seal the terrarium.

Family Involvement

Children may take care of keeping their parents well-informed as to the progress of their seedling experiences. Families can be asked to save eggshells for the project. Their help will be needed in providing a good growing location and in overseeing the care of the seedling that is sent home. Dittoed plant care tips could be fastened to the container. Include a few lines about other aspects of the projects that can help parents become informed listeners to their children.

EXPLORING AT HOME*

Home activities with seeds and plants have a big advantage over school activities. At home with you, your child can watch beginning-to-end happenings. Together, you can watch the whole life cycle of plants, from seed to plant to seed again. If you have use of a small, sunny piece of ground, plant a few dried beans. Let your child keep them watered. Watch for and talk about first sprouts, blossoms, and tiny beans. Eat some of the beans. Save some to dry. Then shell out the beans to plant next year. If you can, bury the vines after they die down. Let them decay and renew the soil for next year's plants.

Windowsill Botany. Anytime you prepare fruits and vegetables for meals, show your child the seeds you find. Soak some citrus seeds in water a few days, then plant a few in a spray can top in potting soil. Put it on a sunny window sill. Let your child give it some drops of water each day and watch patiently. Plant a dried bean, a few lentils, or popcorn seeds in other containers. See which seeds grow faster, which plants live longer. Encourage thinking and talking about what happens.

Bottle Botany. Suspend a fat, single clove of garlic in a small bottle of water, with the pointed tip out of the water. Poke three toothpicks into the middle of the clove to hold the top of the clove out of the water, if the clove is too small to fit the bottle top. Watch daily for fast, dramatic results. Be sure to keep the bottle filled with fresh water. Keep it in a sunny spot.

Leaf Learning. If you get out to a woods with your child, you could start a family leaf collection. *From Little Acorns to Mighty Oaks,* by Rona Bean, is a combination tree identification guidebook and leaf collection album.

RESOURCES

EARTHWORKS GROUP. *50 Simple Things Kids Can Do to Save the Earth.* Kansas City: Andrews & McMeel, 1990.

FURMAN, ERNA. "Plant a Potato—Learn About Life (and Death). *Young Children,* 46 (Nov. 1990), 15–20.

JENSEN, A., FERGUSON, G., & COE, J. *Flowering Plants.* New York: Grossett & Dunlop, 1982. Good botanical background information, including recognition of the functional parts of the plant and wildflowers.

*These suggestions may be duplicated and sent home to families.

LERNER, CAROL. *Moonseed and Mistletoe: A Book of Poisonous Wild Plants.* New York: Morrow, 1988.

PETERSON, ROGER TORY. *Wildflowers.* (Northeastern & North Central America). Boston: Houghton Mifflin, 1986. Color-coded page corners speed identification of specimens in this pocket-size field guide. Paperback.

RUSSELL, HELEN ROSS. *Ten-Minute Field Trips: Using the School Grounds for Environmental Studies.* Chicago: J. G. Ferguson, 1973. Every city-bound teacher should know this book. Nature's ability to triumph over asphalt and concrete permeates the text.

SHOLINSKY, JANE. *Growing Plants from Fruits and Vegetables.* New York: Scholastic Book Services, 1974. A very inexpensive pamphlet with good instructions for growing plants indoors from cuttings and seeds.

SKELSEY, ALICE, & HUCKABY, GLORIA. *Growing Up Green.* New York: Workman, 1973. Excellent precepts for gardening and other plant life projects with children.

ZIM, HERBERT S., & MARTIN, ALEXANDER. *Trees: A Guide to Familiar American Trees.* A Golden Nature Guide: New York: Golden Press, 1987. Inexpensive, comprehensive pocket guidebook.

5

Animal Life

Animals of all sizes and conditions fascinate many children who are eager to watch, touch, and care for creatures. Other children have limited tolerance for anything that creeps, crawls, or nips. The experiences suggested for this chapter can both expand the knowledge of the creature lovers and soften the feelings of anxious children into moderate respect for the useful and beautiful small animals around us. These concepts will be explored:

- There are many kinds of animals
- Animals move in different ways
- Each animal needs its own kind of food
- Many animals make shelters to rear their young

The feasibility of the experiences will depend upon having specimens for observation. Greater flexibility in use is needed for this chapter than for others. It would be sheer luck for a spider and a teacher to meet precisely on the day set aside for spider study.

Buying, housing, and maintaining classroom pets can be expensive. However, earthworms, spiders, and insects exist everywhere and can be easily obtained. The experiences that follow use insects, worms, fish, wild birds, and pictures to illustrate concepts. Suggestions for acquiring and understanding insects, as well as ideas pertaining to borrowed pets, are included.

CONCEPT: There are many kinds of animals

Introduce this topic with a question for children to think about: "What is an animal?" Responses about specific animals will flow easily. Then suggest that there are so many kinds of animals in the world that it would take days just to say their names. "Here is a shorter way to say what an animal is: "*An animal is any living thing that is not a plant*." How many animals can the children think of now? Their list can include people, spiders, earthworms, and insects. There are more than 800,000 species of insects alone! They all have some features in common.

1. What is an insect?

LEARNING OBJECTIVE: To discover ways in which insects are similar to each other.

MATERIALS:

Temporary cages (see page 73)

Live insects*

Preserved insect in plastic box or vial

Magnifying glasses

Paper

Crayons

GETTING READY:

Follow capture techniques below.

Keep live insect cages out of direct sunlight.

SMALL GROUP ACTIVITY:

1. Ground rules for observation should be set: (a) Insects or other small animals stay in the cages and (b) cages must be handled gently.
2. Suggest looking for things that identify members of the insect family: three body parts (head, thorax, abdomen); six legs; two feelers (antennae).
3. Encourage children to try drawing pictures of the insect they are watching.
4. Try to mention: (a) These are adult insects. First they were eggs, then larvae (wingless, wormlike) before changing to adult form. (b) Insects have no bones. They have stiff coverings protecting their soft bodies.
5. Release insects outdoors at the end of the day.

*Spiders are not insects. They have eight legs and are classified as arachnids. Caterpillars are insects. Although they appear to have more than six legs, only six are true, jointed ones.

CAPTURE TECHNIQUES

Locating Creatures

Start looking for specimens near your own doorstep. On warm nights check screen doors for insects that are attracted by light. Hunt for web-building spiders on windowframes and shrubs. Turn over rocks on the ground to find crickets, beetles, and such. Look in flower borders for ladybugs, bumblebees, grasshoppers, ants, and wandering spiders that chase their prey instead of catching them on webs. Examine weed clumps like Queen Anne's lace and milkweed for caterpillars.

Catching Creatures

Cold-blooded animals do not move rapidly in cool parts of the day. A bumblebee is groggy and easy to catch early on a crisp fall morning. Scoop it up with an open jar, then quickly clap on the lid. Slip in a dewy sprig of the plant that the bee rested on, and re-cover the jar with a piece of nylon hosiery or netting, held with a rubber band. Bees may be easier to locate during warmer parts of the day *but quicker to defend themselves.* Try using bamboo toast tongs to catch an alert bee. Use tongs to pick up very spiky-looking caterpillars. Some cause skin rashes if touched.

FIGURE 5–1

Wandering ground spiders and grasshoppers may be caught by clapping an open jar over them. Both species seem to hop straight up inside the jar, making it easy to slip the lid under it.

Nets are preferred for the safe capture of butterflies and moths. Sweep the net over the insect; flip the bag to fold it over the catch. Remove the butterfly by gently holding two wings folded back together. Slip it into a waxed paper sandwich bag. Keep it out of the sun until it can be transferred to an observation terrarium or box.

Only two small spiders have a dangerous bite. Avoid:

- *The black widow:* Shiny black body with bright red hourglass mark beneath the abdomen (back section). Young may have three red dots on top of the abdomen.
- *The brown recluse:* Rare. Lives indoors in attics, closets, or other dark corners. Yellow to brownish color. Fiddle-shaped mark on top of front section (cephalothorax) is dark brown. The base of the fiddle is between the feelers; the neck of the fiddle runs toward the spider's abdomen.

TEMPORARY HOUSING

There are several ways to make inexpensive cages for small creatures. *Making Your Own Nature Museum,* by Ruth McFarlane, shows many of them. However, it is good ecology teaching to release a specimen after a day in the classroom. Comment that the insect is needed outside for cleanup work or pest control, or to help flowers make seeds (see pages 76 and 95). This also skirts the problem of providing live food for some small animals.

A simple temporary cage for small insects can be made by covering a clean plastic tumbler with a piece of nylon hosiery, stretched and held in place with a strip of tape. For larger insects, cut large windows in the sides of a cottage cheese carton, as shown in Figure 5–1. Pull the foot half of an old nylon stocking over the carton, gathering the top tightly with a rubber band. Snap the carton lid over the gathering. Put in a bit of wet cotton for moisture. Keep cages on hand to use whenever children bring small creatures to share. (It can be difficult to sell some

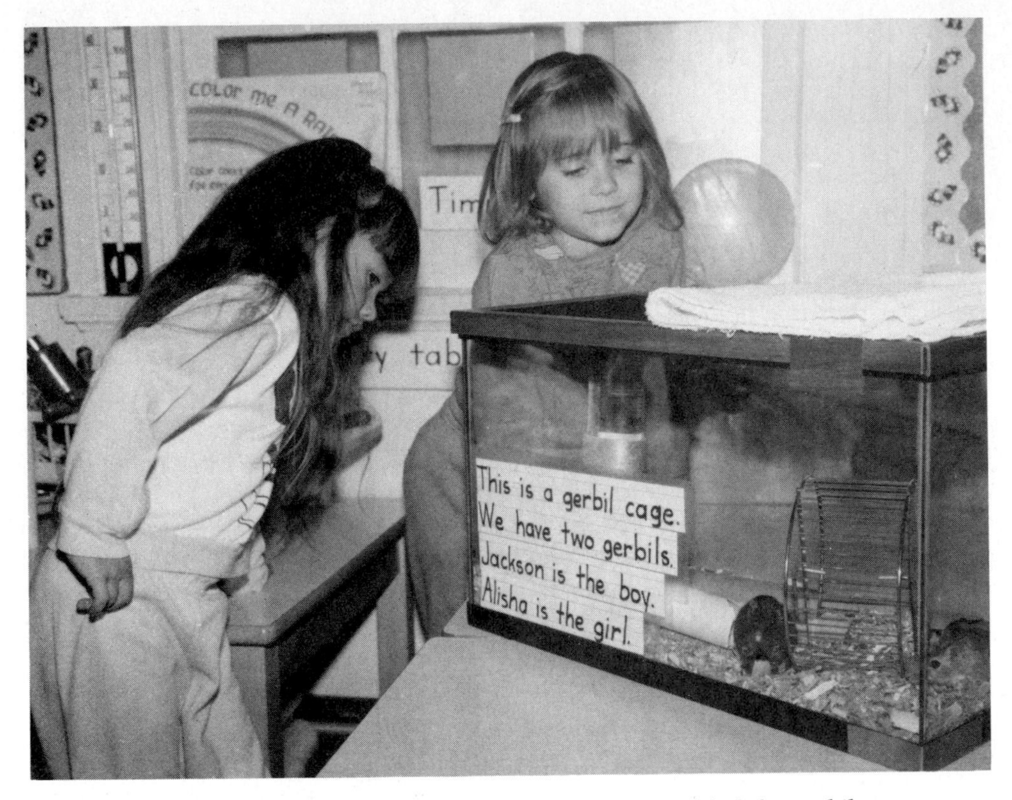

Word recognition comes easily when important facts label the gerbil cage.

children on the cage idea. Joey, for one, was in favor of keeping his caterpillar on a leash of "'gotch tape.")

One fascinating exception to the same-day-release rule is a mother spider. She does not feed during the period of egg-sac construction, nor during the weeks before the spiderlings emerge from the sac. The spiderlings also have a stored food supply for use after their birth.

Wolf spiders fasten egg sacs to their undersides. The babies migrate to the mother's back after emerging. There they cling to knob-tipped hairs as passengers. A wolf spider and her young can survive for a week of observation in a plastic playing-card box if moisture is provided.

A potential disappointment with spider watching is that the mother spider spins an egg sac whether the eggs have been fertilized by a male spider or not. If the eggs are infertile, the mother does not tear the sac open.

PRESERVING SPECIMENS

Youngsters with a strong fear of insects may shrink from examining caged live insects. They may be able to approach a mounted insect that is sealed in a plastic box instead. Start a small collection of insects at opportune moments. Preserve them for children to inspect in school.

Do the preserving away from children. Soak a cotton ball in nail polish remover. Drop it into a small, screw-top jar with the insect. Later glue the specimen to the bottom of a small plastic box. Tape the lid shut. The insect won't skid around inside the box when the children handle it.

Butterfly mounting is not difficult to do. Directions are given in most butterfly study guidebooks.

INSECT PESTS

Some insects do more harm than good to humans and plants. Among them are flies, mosquitoes, cockroaches, black widow and brown recluse spiders, miller moths, and clothes moths. Many insects that sting to defend themselves will not bother people if they are not disturbed. Among them are wasps, hornets, bees, and spiders (Table 5–1).

2. *What is a mammal? (Classifying mammals and nonmammals)*

LEARNING OBJECTIVE: To become aware of mammalian characteristics.

MATERIALS:

Part 1

Caged small mammal:* gerbil, mouse, guinea pig

Part 2

Plastic models or mounted small pictures of mammals (include a human), large and small, domestic and wild, and non-mammalian animals (birds, fish, amphibians, reptiles, insects, spiders)

GETTING READY:

Collect small colored pictures of animals. Mount on index cards, or rubber cement to small yogurt carton lids.

Set up YES/NO trays. Provide fur, picture of nursing animal as YES tray cues. Provide shells, eggshells, reptile skins, or pictures as NO tray cues.

SMALL GROUP ACTIVITY:

1. Establish rules for observing, not disturbing, animal. Guide children to notice mammal characteristics:
   A. "How is the animal covered?" (All mammals have some hair.)
   B. "Why do its sides move in and out?" (All mammals need to breathe air.)
   C. "If we could stroke it, could we feel bones under the fur and skin?" (All mammals have backbones.)
   D. "Do you suppose it hatched from an egg, or did it grow in its mother's body before it was born?" (Mammals develop in their mother's body.)
   E. "How do you think this animal was fed as a baby?" (Female mammals make milk to nurse their young.)
   F. Ask children if people have hair, bones, lungs to breathe; if they developed inside their mothers; if mothers can nurse their babies? (Humans are mammals.)
2. Invite pairs of children to play the mammal classifying game with the pictures or model animals.

*A local pet store or nature center may be a lending resource.

TABLE 5–1 Interesting, helpful insects

INSECT	FUNCTION	INTERESTING FEATURES
Ants	Scavengers who clean their surroundings. Newly discovered function: pollinators of flowers.	Social insects who live in colonies with separate jobs to perform; some nurse young, some just gather food.
Bees	Highly valued pollinators of plants. Producers of honey and beeswax.	Social insects who live in colonies with specialized jobs to perform.
Butterflies	Pollinators of plants. (Explain to children that they help plants make seeds.)	Slender bodies. Antennae have ball tips. Fly by day. Fold wings straight up when resting. Usually form a chrysalis.
Moths	Some are pollinators. Silk moths spin strong, lustrous fibers that are made into fabric.	Fat, furry bodies. Feathery antennae. Wings spread flat when resting. Usually spin cocoons. Fly at night.
Beetles	Some kinds are scavengers who tidy up. Ladybug beetles are valued for pest control by gardeners. Some beetles destroy crops.	Many handsome varieties: striped, spotted, iridescent.
Crickets	Serve as food for other animals. They are destructive to some crops.	Only male crickets chirp. They raise and rub their hard wing covers to make vibrations.
Fireflies (soft-bodied beetles)	Appreciated for adding charm to summer nights.	Fireflies signal to one another with flashes of light.
Grasshoppers, Locusts	Destroy some crops but serve as food for other animals.	Wings barely visible when flying. Jump with hind pair of legs. Hearing area is in the abdomen.
Praying Mantis	So valuable for pest control that egg cases are sold to gardeners.	Almost 4" (10 cm) long, so body parts easily seen. Frightening to see them tearing other insects apart.
Wasps and Hornets	Some wasps pollinate fruit trees. Some eat destructive larvae.	Hornets and some wasps chew dead wood into paper to make nests for their young. Some wasps make nests from mud. Beautiful engineering.

CONCEPT: Animals move in different ways

1. How does an earthworm move?

LEARNING OBJECTIVE: To observe and describe the ways in which different animals move.

Introduction: "Do you need six legs to walk on? If your body were bent over and close to the ground, would more legs help keep you balanced? What helps some insects and birds move through the air? Can you think of a very small animal that has no legs—one that only has muscles and tiny bristles to help him move? Perhaps we'll find some worms to watch when we go outdoors."

MATERIALS:

Shovels and a digging place

Produce trays

Light-colored sand

Bucket, stick

Soil

Peat moss

Water

Newspapers

Plastic shoe box

Nylon net, rubber band

Dry oatmeal, coffee grounds, or cornmeal

GETTING READY:

Check soil—where you have permission to dig—for the presence of earthworms before taking children. If children can't do the digging, bring worms and a shovelful of soil to school in a bucket. Cover with a plastic bag. See *How to Hold an Earthworm,* in the following section.

SMALL GROUP ACTIVITY:

1. Let children dig for earthworms, if possible. Give them equal amounts of peat moss and soil to mix with a stick in the bucket. Add water to dampen well.
2. Spread a thin layer of sand on bottom of the shoe box. Cover with 3" (8 cm) of mixed soil.
3. Sprinkle meal or coffee grounds on top; scratch in with the stick. Put all but two earthworms on the soil. Cover top with net fastened with large rubber band.
4. Put two earthworms on trays to watch them move. (Bristles are retractable and hard to see. A wide, light band at midsection is the egg case. The tail end is tapered and the head end is rounded. They have no eyes.)
5. See the following section for earthworm farm care. Check sides and bottom for signs of change. Dark soil streaks in the light sand show how earthworms mix soils.

CARE OF EARTHWORMS

Earthworm Farm Care

Wrap black paper around the sides of the shoe box to encourage tunneling where it can be easily seen. It will take a while for the earthworms to get used

A fish, a boy, and a hand lens come together for quiet observation of fins in motion.

to the new soil before they start tunneling. Remove the paper for short periods of time to check for signs of activity.

Earthworms will not survive long in dry soil. Keep the farm away from a heat source. Sprinkle the soil frequently to keep it damp. The box lid can be placed loosely over the net to retain moisture, but some air must enter the box.

Work in a spoonful of dry food each week. Try adding bits of grass and leaves. Tiny balls surrounding tunnel holes on the surface of the soil are the castings of soil digested by the earthworms. This is one way that worms enrich the soil. They also help water and air reach plant roots as they make their tunnels.

How to Hold an Earthworm

Youngsters credit adults with unlimited ability and courage. Measuring up to this idealism can be hard for some of us when it means holding a child's cherished earthworm. Bolster your courage for this eventuality with some understanding about your tactile senses: compare the sensation of holding a slippery object in the palm of the hand with that of holding the same object between the thumb and forefinger. The palm of the hand has fewer nerve endings to convey tactile sensations. Therefore, a placid worm that is loosely cupped in the palm of the hand can scarcely be felt.

2. *How does a fish move?*

LEARNING OBJECTIVE: To observe and describe the ways in which different animals move.

MATERIALS:

1 widemouthed gallon jar

Aquarium gravel or lake sand

Rocks

Water plant—purchased or from lake or stream

1 small goldfish

Black paper

Newspaper

Water

GETTING READY:

Involve children in as many preparation steps as you can.

Let a gallon of water stand overnight in open containers (so chlorine can escape).

Wash local water plants carefully. Wash lake sand in a deep pan by letting a slow stream of *hot* water fill pan. (Plants and gravel from a good pet store need not be washed.)

Put an inch of sand or gravel in jar. Put plant roots in sand and anchor with a rock.

Cover sand with folded newspaper while pouring in water. Remove paper.

Put fish in water.

SMALL GROUP ACTIVITY:

1. Look at the body of the fish. Discuss how it is different from the earthworm's body.
2. Recall the looping, sliding movements of worms. Compare with varied movements of the fish: it darts up, down, forward, backward, or rests.
3. Find 7 fins: 2 pairs approximately where our arms and legs grow, topside, underside, and tail fins. (Arm and leg position fins work fast to move fish forward and back, top and bottom fins give balance, tail fins steer fish as it swings from side to side. Some fins look like fancy decorations, but they have delicate bones, and they work hard.)
4. Tap the tank lightly. Does the fish go faster? Does it move differently to speed up? Watch closely.
5. To keep the fish as a class pet, put fish feeding on a rotating routine chart. Each day a different child feeds it a *pinch* of tropical fish food. No weekend feeding is needed. Keep the aquarium away from direct sunlight. Tape dark colored paper around the side closest to the window to cut down algae growth. Some light is needed. Siphon or dip out one third of the water; replace with aerated water as needed.

CONCEPT: Each animal needs its own kind of food

Introduction: "Did you have a nice bowl of acorns and a plateful of grass with ladybugs for breakfast today? Why not? What did you have?" Help children recognize that each kind of animal needs its own kind of food in order to live. Read *Cow's Party,* by Marie Hall Ets. Discuss making a feeder to help winter birds get the kind of food they need. Plan to maintain the feeder until spring. Some birds may come to depend on that food supply and starve without it.

1. Can we feed winter birds?

LEARNING OBJECTIVE: To value helping wild animals find food.

MATERIALS:	SMALL GROUP ACTIVITY:
	A
Gallon milk carton or clean plastic bleach jug 12" (30 cm) stick Scissors Twine Commercial wild bird seed, plus seeds saved from plant experiences	Cut out two sides of carton or jug, leaving a two-inch border near the bottom to hold seeds. Poke the stick through the feeder in border edges to form two perches. Pierce holes through the carton top for hanging twine, or wind twine around jug top and handle. Tie to a tree branch low enough for easy refilling.
	B
Dried pinecones String Peanut butter	Wind string around top scales of cones; make a hanging loop. Let children stuff peanut butter between scales with spoon handles.
	C
Half-pint milk cartons, tops off String Melted suet Seeds Spoons, bowl	Melt suet away from children. Use a heavy, deep saucepan over low heat. When cool, divide into bowls. Let children mix cooled suet with seeds, then fill cartons with mixture.
	D
Pint plastic berry baskets Yarn Suet	Let children lace two baskets together with yarn. Add chunks of suet before last side is lashed together. Tie to tree trunks.

Note: Locate feeders out of the reach of squirrels or cats. Sprinkle seeds on the ground beneath feeders, when they are put out, to attract birds.

CONCEPT: Many animals make shelters to rear their young

1. Do ants care for their young in nests?

LEARNING OBJECTIVE: To observe the social nature of ants.

Introduction: "When you were a tiny baby, did your mother put you outdoors to find your own food? Why not? Many kinds of animals make shelters where they can take care of their babies. Perhaps we can see how ants do this." Discuss making an ant farm. If this cannot be done, try to observe an anthill outdoors. Gently poke into the hill with a stick. The children may be able to see nurse ants carrying eggs, larvae, and cocoons as they leave their disturbed nest.

MATERIALS:

Widemouthed quart jar

Slim olive jar to fit in the quart jar

Construction paper

Rubber bands

Piece of nylon hosiery

Bit of sponge

Flashlight

Jam or syrup

Newspapers, trowel, double grocery bag

Ants, nest soil

GETTING READY:

Do this away from children the first time. *You may be very busy for a while.* Locate and dig into nest of medium or large ants. Quickly scoop ants, larvae, eggs, cocoons, and soil into bag. Fasten top well. (Do not mix ants or soil from two nests.)

Keep bag in cool place until needed.

SMALL GROUP ACTIVITY:

1. Make this ant farm outdoors. Let children center closed olive jar inside the quart jar; fold and tuck newspaper into top of quart jar as a funnel.
2. Gently transfer ants and soil from bag to jar. Quickly stretch hosiery over jar top; slip on a rubber band. Let children wet the sponge and place a dab of jam on it. Quickly slip it into the jar and cover again with hosiery.
3. Let children wrap black paper around the jar, fastening it with rubber bands or tape.
4. Let children keep the sponge damp by dropping water through the hosiery daily; replenish jam weekly.
5. In a few days let children remove paper for a few minutes to see what the ants are doing. A flashlight may be helpful. Lift jar to check activity at the bottom of the jar. Rewrap jar after viewing. Try to see food gatherers, nurses for young, nest cleaners, soldiers, and hump-backed queen.

2. *Can we help nest-building birds?*

LEARNING OBJECTIVE: To value helping wild birds provide shelter for rearing their young.

Introduction: "Do you think that a mother bird lays eggs on tree branches and leaves them to hatch by themselves? Of course not! She works hard with the father bird to build a nest where she can keep the eggs warm and keep the hatched babies safe and fed. When the babies are big enough to take care of themselves, the whole family leaves the nest and never uses it again. Let's take a close look at a nest." (Hunt for your nest when deciduous trees have lost their leaves. Look in dense shrubs, heavy vines, tree crotches, and under eaves. Ask your county agricultural extension agent for suggestions on finding nests. Check for exhibits in a library or museum, if you cannot find one.)

MATERIALS:

Abandoned bird nest

Clear plastic shoe box

Tape

Plastic berry baskets

6″ (15 cm) pieces of yarn

Excelsior

Dried tall grass

SMALL GROUP ACTIVITY:

1. Let children examine the nest structure. "Is it lined with special material? Why?" Keep nest in the plastic shoe box, taped shut for easy handling and storage.
2. If another nest is available, let children gently pull it apart to see how hard the birds work, piece by piece, to build nests. Try to rebuild it.
3. Let children prepare nest materials for birds that come to school feeders. Pull apart clumps of excelsior, place loosely in berry baskets, and work some strands through the holes. Weave strands of yarn and long grass through bottom and side holes of other baskets.
4. Hang baskets near feeders.

Read *No Roses For Harry,* by Gene Zion.

ANIMALS IN THE CLASSROOM

The Borrowed Pet

Much can be learned by children who help provide the daily life requirements of a classroom pet. Many of those learnings can also be sampled during a short visit by a borrowed pet.

Before the pet arrives, discuss with the children safe ways to watch and care for it. For small pets accustomed to pens, make an observation box from a large carton. Cut windows in the sides and cover them with plastic wrap, if needed. Some pets are better off being held by their owners during the visit. Try to let children see the animals eating and drinking water. Help children find answers to questions about how the animal moves, gets its food, and protects itself. Does the pet have bones inside, or a hard outside covering? Does it have hair or fur, smooth or feather-covered skin? Does it nurse its young? Does it build a shelter for its young?

Recent animal visitors to our school included a pet boa constrictor, who laced himself through the rungs of a chair, and a small pony who was unexpectedly led into the building by a mischievous owner.

Animal Rearing

Rearing butterflies or moths from caterpillars, chrysalid, or cocoon stages can be enthralling or disappointing. Try following the procedures described in *Terry and the Caterpillars,* by Millicent Selsam. Strong interest and luck are required for success. (Childcraft sells a Butterfly Garden with caterpillars.)

The same mixture of dedication and luck also contributes to a good outcome with an egg-incubating project. A commercial incubator is probably a better choice than improvised equipment for classroom use. Follow the instruc-

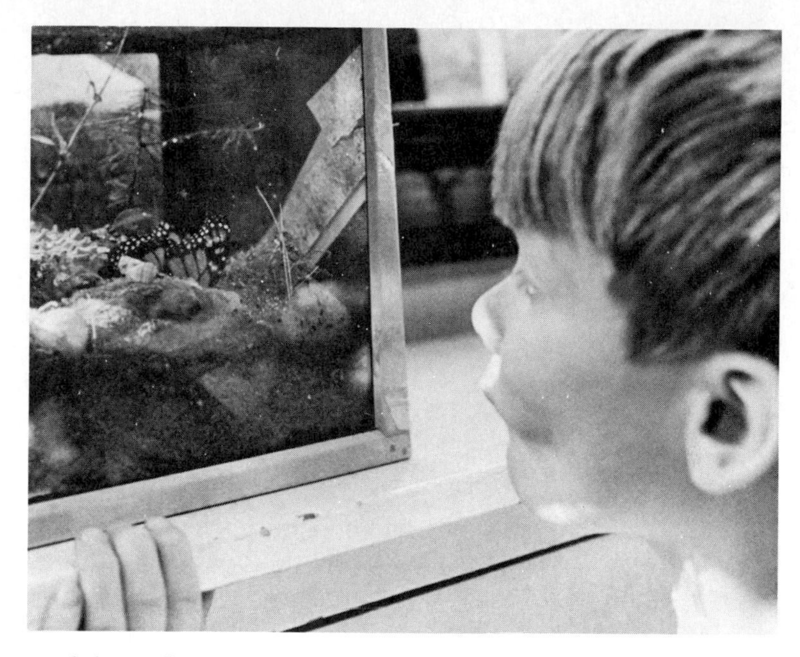

A monarch butterfly rested in our terrarium, inches away from awed observers.

tions included with the incubator. Be sure the fertile eggs have not been allowed to cool after being laid. Plan to find a home for the chick or duckling after it has hatched.

Are Classroom Pets Necessary?

Good teachers allow for individual differences in children's responses to animals. They do not assume that all children adore animals. Also, they do not insist that a fearful child make physical contact with animals.

Teachers should extend the same consideration to their own feelings about animals. While they should avoid expressing negative attitudes about animals to children, they need not feel obligated to undertake yearlong care if they do not enjoy the experience. Neither should they feel inadequate if they prefer to skip the responsibility of animal care. The teacher's primary affective focus and primary responsibility are caring about and caring for children. Feeling similar warmth and commitment toward animals is an asset, but not a requirement.

PRIMARY INTEGRATING ACTIVITIES

Music (Resources in Appendix 1)

Many traditional and contemporary children's songs complement ongoing animal life study:

"Shoo Fly, Don't Bother Me"
"The Eency Weency Spider"
"Jig Along Home," by Woodie Guthrie (Worms and grasshoppers are among the dancers.)
"Three Little Ducklings"

A Malvina Reynolds' song, "You Can't Make a Turtle Come Out," reminds children that to observe animals ". . . . you'll have to patiently wait."

Listen to *Birds, Beasts, Bugs and Little Fishes,* sung by Pete Seeger. The songs were written by Ruth Seeger in *Animal Folk Songs for Children.* Also listen to "I Like the Animals In the Zoo," sung by Ella Jenkins on *Seasons for Singing.* The song provides a good pattern for improvising about the animals that visit your room:

I like the grasshopper, in the jar.
I like the grasshopper, it jumps far.

Animal horns were among the earliest forms of musical instruments. Conch shells are still used as horns in some parts of the world. One of the percussion instruments on our classroom music shelf is an unoccupied box turtle shell. The children enjoy playing it.

Math Experiences

Many commercial math materials incorporate animals:

1. Animal Match ups: Puzzle cards with sets of animals designed to link with numerals.
2. Magnet Board Forms: Plastic ducks and rabbits are used in magnet counting sets.
3. "See-Quees" puzzles include the development from egg to butterfly; from egg to frog; and from nest building to robin egg hatching. The pictures are to be arranged in frames according to time sequence.

Use collections of small seashells as materials for counting, classifying, set making, or numeral game markers.

Keep a flannel board tally of small animals brought into your classroom or observed on walks. Mount pictures of the animals cut from magazines on sandpaper or flock-backed adhesive paper backing. Bring out the envelope of pictures for children to count when new observations are reported.

If your class tries an incubating project, make a chain of 21 large paper clips to represent the days of the incubation period. Remove a clip each day and count the days left in the waiting period.

Stories and Resources About Animal Characteristics

ATTMORE, STEPHEN. *Animal Encyclopedia.* New York: Checkerboard Press, 1989. An illustrated general reference organized by the classes of the animal kingdom: mammals, reptiles, amphibians, birds, fish and insects.

COLE, JOANNA. *Large as Life: Daytime Animals.* New York: Alfred A. Knopf, 1985.

______. *Large as Life: Nighttime Animals.* New York: Alfred A. Knopf, 1985. These companion books feature life-size paintings of wild animals. The beautifully rendered textures of fur, feathers and spines are outstanding. The text describes animal characteristics of interest to young children. Nature notes at the end provide supplementary information.

EPSTEIN, SAM & BERYL. *Bugs For Dinner? The Eating Habits of Neighborhood Creatures.* New York: Macmillan, 1989. Menus and more about insects and small animals familiar to city children.

GARDNER, BEAU. *Guess What.* New York: Lothrop, 1985. Animal silhouettes are presented on one page, for children to observe and predict the animal, and shown in detail on the following page.

LILLY, KENNETH. *Animal Builders.* New York: Random House, 1984. Also in the series: *Animal Climbers, Animal Jumpers, Animal Runners,* and *Animal Swimmers.* Simple concepts about animal characteristics for the youngest preschoolers, printed on durable board for easy handling.

WHITCOMBE, BOBBIE. *Animals in Danger.* New York: Checkerboard Press, 1990. Provides clearly written capsules of information in large type for young readers. Intriguing illustrations are generously used. Also in this series of inexpensive, hardcover books are: *Animals of the Night, Dangerous Animals,* and *Strange Animals.*

Stories and Resources About Insects, Spiders, and Earthworms

BRANDT, KEITH. *Insects.* Mahwah, NJ: Troll Associates, 1985. Interesting information promotes respect for these small creatures. Fresh perspectives on moving insects in Robin Brickman's illustrations. Paperback.*

CARLE, ERIC. *The Very Busy Spider.* New York: Philomel Books, 1985. A spider's workday is revealed in this engaging story for preschoolers.

CAUDILL, REBECCA. *A Pocketful of Cricket.* New York: Holt, Rinehart and Winston, 1964. A cricket brought from home helps ease a boy's first days at school.

CORNELIUS, CAROL. *Isabella Wooly Bear Tiger Moth.* Chicago: Children's Press, 1978. How one caterpillar hatches, survives a hazardous life, and metamorphoses into a yellow moth.

CRAIG, JANET. *Amazing World of Spiders.* Mahwah, NJ: Troll Associates, 1990. Fascinating information about the diversity of spiders and the intricacy of their work. Paperback.*

DORROS, ARTHUR. *Ant Cities.* New York: Thomas Y. Crowell, 1987. The organized life beneath the anthills is carefully presented for young children.

FISCHER-NAGEL, HEIDEROSE. *Life of the Ladybug.* Minneapolis: Carolrhoda Books, 1986. Fascinating photographs of the life cycle of ladybugs from egg laying and hatching, to metamorphosis and hibernation.

FLORIAN, DOUGLAS. *Discovering Butterflies.* New York: Charles Scribner's Sons, 1986. Well-paced information about butterfly life, with a good balance between facts and illustrations, make this an excellent resource for preschool, primary grade children.*

FRISKEY, MARGARET. *Johnny and the Monarch.* Chicago: Children's Press, 1961. Johnny's duck hatches a brood during the span of a butterfly's life cycle.

GRAHAM, MARGARET BLOY. *Be Nice to Spiders.* New York: Harper & Row, 1967. A fine story about a garden spider that improves life for the animals in a zoo. It promotes respect for spiders.

HOGNER, DOROTHY CHILDS. *Earthworms.* New York: Thomas Y. Crowell, 1953. Comprehensive information.

*Starred references, written at the young child's level of understanding, can help teachers with minimal backgrounds in science expand their knowledge base.

MOUND, LAURENCE. *Insect.* New York: Alfred Knopf, 1990. Beautifully organized information and sparkling color photographs. Eyewitness Series.

NORSGAARD, E. JAEDIKER. *How To Raise Butterflies.* New York: Dodd, Mead, 1988. Exact color photographs enrich this book that takes the guesswork out of rearing butterflies in the classroom.*

PARSONS, ALEXANDRA. *Amazing Spiders.* New York: Alfred A. Knopf, 1990. Children will become better observers because of the simple text and exceptional illustrations in this Eyewitness Junior book. Paperback.

PODENDORF, ILLA. *The New True Book about Insects.* Chicago: Children's Press, 1981. Excellent close-up color photographs of insects. Easy text for young readers.

POLLOCK, PENNY. *The Spit Bug Who Couldn't Spit.* New York: G.P. Putnam's Sons, 1982. A factually based tale of a lonely spittlebug who learns to cope.

SABIN, FRANCENE. *Amazing World of Ants.* Mahwah, NJ: Troll. 1982. Information about familiar and rare species of ants, written in a comfortable style for reading aloud. Paperback.*

SELSAM, MILLICENT. *A First Look At Caterpillars.* New York: Walker, 1987. When closely observed, movements and markings, horns and hairs become the interesting criteria for identifying varieties of caterpillars.

VAN ALLSBURG, CHRIS. *Two Bad Ants.* Boston: Houghton Mifflin, 1988. This imaginative story of adventuresome ants can spark children's creative thinking and perspective-taking.

WHALLEY, PAUL. *Butterfly and Moth.* New York: Alfred A. Knopf, 1988. Beautiful illustrations and clear information will draw children to this book.

WHITCOMBE, BOBBIE. *Insects.* New York: Checkerboard Press, 1990. Fascinating facts and carefully detailed illustrations invite young readers.

Stories and Resources About Fish, Mollusks, Amphibians, Reptiles

BERNARD, GEORGE. *Common Frog.* New York: G.P. Putnam's Sons, 1979. Excellent color photographs depict the three years of development from egg to full-grown frog.*

CARLE, ERIC. *A House For Hermit Crab.* Natick, MA: Picture Book Studio, 1988. A wonderful blend of accurate portrayal of ocean floor life with a metaphor about the pangs of growth and change in children's lives.

COLE, JOANNA. *Hungry, Hungry Sharks.* New York: Random House, 1986. Describes many kinds of sharks, from hand-size dwarfs to giants. Reassures that sharks do not go hunting for people. Paperback.

DALY, KATHLEEN. *The Golden Book of Sharks and Whales.* New York: Golden Books, 1989. Sharks, the biggest fish, and whales, the biggest mammals, seem a bit less fearsome with this interesting information and graceful art.

DAVIS, ALICE V. *Timothy Turtle.* New York: Harcourt, Brace, 1940. Clear thinking solves a turtle's problem when misguided group effort fails.

FINE, JOHN. *Creatures of the Sea.* New York: Atheneum, 1989. Fascinating color portraits and descriptions of unusual sea dwellers.

FLORIAN, DOUGLAS. *Discovering Frogs.* New York: Charles Scribner's Sons, 1986. Presents a good balance between illustrations and facts that will interest preschool and primary grade children.

______. *Discovering Seashells.* New York: Charles Scribner's Sons, 1986. Offers good information about sea mollusks for the preschool and primary grade level.*

GOUDEY, ALICE. *Houses From the Sea.* New York: Charles Scribner's Sons, 1959. Both a story and a reference book, with lovely illustrations by Adrienne Adams.

KOHN, BERNICE. *The Beachcomber's Book.* New York: Viking Press, 1987. Describes many of nature's gifts from the sea, and how to use them in children's crafts and cooking.

McCloskey, Robert. *One Morning in Maine.* New York: Viking Press, 1952. Clam digging and shore animal life are part of this classic story.

Myers, Arthur. *Sea Creatures Do Amazing Things.* New York: Random House, 1981. Fascinating information about sea life from microscopic creatures to the largest mollusk in the world.

Parker, Nancy. *Frogs, Toads, Lizards and Salamanders.* New York: Greenwillow, 1990. Whimsical jingles and illustrations accompany the scientific descriptions and drawings of amphibians. A good retention device.

Parsons, Alexandra. *Amazing Snakes.* New York: Alfred A. Knopf, 1990. Vivid illustrations and a text designed to intrigue young readers. Paperback.

Ryder, Joanne. *Snail in the Woods.* New York: Harper & Row, 1979. A satisfying story that reveals the nature of the common snail.

Sabin, Louis. *Reptiles and Amphibians.* Mahwah, NJ: Troll, 1985. A good source of answers for older preschoolers. Paperback.

Selsam, Millicent. *Plenty of Fish.* New York: Harper, 1960. An enjoyable story that informs about life requirements of fish in an aquarium.

Smith, Trevor. *Amazing Lizards.* New York: Alfred A. Knopf, 1990. Vivid illustrations and a text designed for young readers. Eyewitness Juniors.

Ungerer, Tomi. *Crictor.* New York: Scholastic Book Services, 1971. An engaging adventure story for preschoolers about a boa constrictor.

Whitcombe, Bobbie. *Whales and Sharks.* New York: Checkerboard Press, 1990. Generous illustrations, large type to engage young readers.

Stories and Resources About Birds

Burnie, David. *Bird.* New York: Alfred A. Knopf, 1988. Beautifully organized information and sparkling color photographs. London Natural History Museum Eyewitness series.

Curran, Eileen. *Bird's Nests.* Mahwah, NJ: Troll, 1985. The very simple text is intended for preschoolers, but the lovely, precise illustrations will catch the attention of older children also.

Day, Jennifer. *What Is A Bird?* New York: Golden Books, 1975. Organized according to classes of birds, this book teaches bird characteristics through catchy repetition. Beautifully illustrated. Paperback.*

Flack, Marjorie, & Wiese, Kurt. *The Story About Ping.* New York: Viking Press, 1961. Ping, a duck, wanders away and returns to his houseboat home on the Yangtze River in China.

Gans, Roma. *When Birds Change Their Feathers.* New York: Thomas Y. Crowell, 1980. Fascinating feather facts.

Goldin, Augusta. *Ducks Don't Get Wet.* New York: Harper & Row, 1989. Good information, an experiment to explain the title concept, and fine prose in this revised classic. Paperback. Free activity card for teachers.

Kaufman, John. *Birds Are Flying.* New York: Thomas Y. Crowell, 1979. Interesting information about how birds are able to fly.

Parsons, Alexandra. *Amazing Birds.* New York: Alfred A. Knopf, 1990. Outstanding color photographs, clear and easy text in this child's version of the London Natural History Museum Eyewitness Series. Paperback.

Santrey, Laurence. *Birds.* Mahwah, NJ: Troll, 1985. A brief, general reference covering common characteristics, nesting, and migration. Paperback.

Zion, Gene. *No Roses For Harry.* New York: Scholastic Book Services, 1971. Nesting birds and Harry's detested sweater come together.

Stories and Resources About Mammals

Bernard, George. *Harvest Mouse.* New York: G.P. Putnam's Sons, 1981. Beautiful photographs reveal the dangerous life of one of these tiny mammals and her babies.

GOODALL, JANE. *Chimps.* New York: Byron Press Books, Macmillan, 1989. Written to promote respect and understanding for animals and the habitat mankind must preserve for them, these Jane Goodall's Animal World books will appeal to older primary-grade children. Paperback. Other books in this series are: *Hippos, Lions,* and *Pandas.*

HANSEN, ROSANNA. *Horses and Ponies.* New York: Golden Books, 1988. Differences between horses and ponies, breeds and their history are described.

HELLER, RUTH. *Animals Born Alive and Well.* New York: Grosset & Dunlap, 1982. The text is simple and informative. The pages are handsomely designed and illustrated by the author.

HOFFMAN, MARY. *Animals in the Wild: Elephant.* New York: Random House, 1985. Beautiful color photographs and simple information about young and mature animals in their natural surroundings. Paperback. Other books in this series are: *Animals in the Wild: Monkey, Animals in the Wild: Panda,* and *Animals in the Wild: Tiger.*

KAUFMAN, JOE. *Big Book About Mammals and Birds.* New York: Golden Books, 1989. Accurate information and cheerful illustrations are in abundance here. The front and back covers pose clever creative-thinking challenges.

KOMORI, ATSUSHI. *Animal Mothers.* New York: Putnam, 1983. Simple text and lovely illustrations for young preschoolers.

LAVIES, BIANCA. *It's An Armadillo.* New York: E.P. Dutton, 1989. Armadillos can be fascinating and appealing when photographed at close range. Charmingly described by this patient photo journalist-author.

PARKER, STEVE. *Mammal.* New York: Alfred A. Knopf, 1989. Outstanding illustrations and well-organized information in this London Museum of Natural History Eyewitness Series book.

SABIN, FRANCENE. *Mammals.* Mahwah, NJ: Troll, 1985. Because it condenses a wide range of information into a brief book, this reference is more suitable for primary-grade children. Paperback.*

______. *Whales and Dolphins.* Mahwah, NJ: Troll, 1985. Presents corrective information about so-called "killer whales," the friendly, intelligent dolphin, and other seagoing mammals. Paperback.

STEIN, SARA. *Mouse.* San Diego: Harcourt, Brace, Jovanovich, 1985. This highly recommended book combines a warm, lean text with charming, accurate illustrations by Manuel Garcia. It reveals how mice nest, reproduce, and develop prenatally.

Stories and Resources About Habitats

BAYLOR, BYRD. *Desert Voices.* New York: Charles Scribner's Sons, 1981. Lovely prose and sketches of desert-habitat animals.

CURRAN, EILEEN. *Life in the Pond.* Mahwah, NJ: Troll, 1985. A simple text for young preschoolers to enjoy, with vivid illustrations by Elizabeth Ellis. Paperback.

______. *Life in the Forest.* Mahwah, NJ: Troll, 1985. Young preschoolers will enjoy locating forest animals hiding in nests and trees. Paperback.

______. *Life in the Sea.* Mahwah, NJ: Troll, 1985. Beautiful and lively illustrations by Joel Snyder make this book special. Paperback.

EUGENE, TONI. *Creatures of the Woods.* Washington, DC: National Geographic Society, 1985. The color photographs forming the core of this book are intimate and dramatic. A simple text is supplemented with more detailed information for older children and adults.

HELLER, RUTH. *How to Hide a Whip-poor-will.* New York: Grossett & Dunlap, 1986. Stunning illustrations by the author reveal nature's protective coloration, body forms, and behaviors. Paperback.*

HESS, LILO. *Secrets in the Meadow.* New York: Charles Scribner's Sons, 1986. The complex animal life that exists in a country meadow is traced from dawn through the night. Intimate black-and-white photographs.

RINARD, JUDITH. *The World Beneath Your Feet.* Washington, DC: National Geographic Society, 1985. Remarkable, delicate photographs, such as six ladybugs on a dandelion blossom, plus a simple text and supplementary information.*

RYDER, JOANNE. *Animals in the Wild.* New York: Golden Books, 1987. A nice presentation of seasonal activities of familiar woodland animals. Paperback.

______. *Step into the Night.* New York: Four Winds, 1988. The natural world at night is exquisitely portrayed in poetic text. A girl listens to, watches, and imaginatively becomes different nocturnal animals.

______. *Mockingbird Morning.* New York: Macmillan, 1989. A nature-loving poet and an outstanding illustrator combine talents in this lovely child-view of animal life around her home.

SABIN, FRANCENE. *Wonders of the Forest.* Mahwah, NJ: Troll, 1982. Seasonal patterns of forest animals are described with gentle, realistic illustrations by Michael Willard. Paperback.

______. *Wonders of the Pond.* Mahwah, NJ: Troll, 1985. Explores animal life in the pond, from amoeba to muskrats. Watercolor illustrations by Leigh Grant. Paperback.

SABIN, LOUIS. *Wonders of the Desert.* Mahwah, NJ: Troll, 1985. Information about animal protection and survival mechanisms needed for desert life. Paperback.

______. *Wonders of the Sea.* Mahwah, NJ: Troll, 1985. Describes sea animals from the microscopic to the gigantic, including the gentle, giant octopus. Paperback.

SELBERG, INGRID. *Nature's Hidden World.* New York: Putnam, 1983. The simple, carefully written text about woodland animals is enhanced by charming habitat illustrations and intriguing pop-up features.

SELSAM, MILLICENT, & HUNT, JOYCE. *Keep Looking!* New York: Macmillan, 1989. Beautiful in both words and watercolors, this fine book encourages the patient observation of many forms of animal life in a winter backyard.

WOOD, JOHN. *Nature Hide and Seek: Oceans.* New York: Alfred A. Knopf, 1985. An intriguing foldout format demonstrates well the many forms of camouflage fish use to survive in several ocean environments.*

Stories and Resources About Baby Animals, Hatching, and Pets

BURCH, ROBERT. *Joey's Cat.* New York: Viking Press, 1969. A pet-keeping problem is resolved by a middle class family.

COHEN, MIRIAM. *Best Friends.* New York: Macmillan, 1971. The friendship of Jim and Paul is cemented by their mutual rescue of the classroom incubation project.

______. *When Will I Read?* New York: William Morrow, 1977. Danny's concern for the classroom hamsters helps him achieve his goal of becoming a reader.

ETS, MARIE HALL. *Play With Me.* New York: Viking Press, 1955. A girl discovers how to make friends with animals. Good to read before a pet visits the classroom.

HALL, DEREK. *Elephant Bathes.* New York: Alfred A. Knopf, 1985. Also in this series, *Gorilla Builds,* and *Polar Bear Leaps.* Simple stories for preschoolers about young animals learning to cope with their natural environments.

HELLER, RUTH. *Chickens Aren't the Only Ones.* New York: Grosset & Dunlap, 1981. The simple and direct information in the text is vividly reinforced with the author's striking illustrations. A Children's Science Book Award winner.*

INGOGLIA, GINA. *Nature Babies.* New York: Golden Books, 1989. The simple text in this small book offers beginning information about habitat adaptation to young learners. Paperback.

KEATS, EZRA JACK. *Pet Show.* New York: Macmillan, 1974.

MANUSHKIN, FRAN. *Puppies and Kittens.* New York: Golden Books, 1989. This small book about young pets nursing and developing characteristic behaviors is well written for young learners. Paperback.

MCCLOSKEY, ROBERT. *Blueberries for Sal.* Seafarer ed. New York: Viking Press, 1968. Four sets of mothers and their offspring are part of this story: Sal and her mother, plus bear, quail, and crow families.

______. *Make Way For Ducklings.* Seafarer ed. New York: Viking Press, 1969. Finding a suitable home and raising a brood of ducklings occupy Mrs. Mallard's time.

RINARD, JUDITH. *Helping Our Animal Friends.* Washington, DC: National Geographic Society, 1985. Responsible care for pets by children, and professional care of orphaned wild animals.

STEIN, SARA. *Cat.* San Diego: Harcourt Brace Jovanovich, 1985. With simplicity and precision, the author focuses our attention on the predatory, self-sufficient nature of the family cat. The excellent text is enhanced by Manuel Garcia's clever illustrations.

______. *Mouse.* San Diego: Harcourt, Brace, Jovanovich, 1985. This highly recommended book combines a warm, lean text with charming, accurate illustrations by Manuel Garcia. It reveals how mice reproduce and grow.

WARD, LYND. *The Biggest Bear.* New York: Scholastic Book Services, 1975. A touching lesson about trying to domesticate a woods animal.

Stories and Resources About the Death of Pets

ROGERS, FRED. *Mister Rogers First Experience Book: When a Pet Dies.* New York: Putnam, 1988. Mister Rogers helps children manage their feelings of loss, loneliness, and frustration after a pet dies. Highly recommended.

VIORST, JUDITH. *The Tenth Good Thing About Barney.* Atheneum, 1971. The conclusion of this story may not be acceptable to some. Decide for yourself.

Poems (Resources in Appendix 1)

"The Little Land" from *A Child's Garden of Verses,* by Robert Louis Stevenson, records a charming perspective of insects projected by a diminutive sailor on a leaf boat. A newly hatched turtle experiences the life in the field around him in the charming collection of poems, "Inside Turtle's Shell," by Joanne Ryder.

Many other poets create delicate or amusing sketches of insects and somewhat larger animals. Poems are found in these collections:

Brown, Margaret Wise, *Nibble Nibble:*
"A Child's Delight" (a firefly in a jar)
"Bumble Bee"

de Regniers, Moore, and White, *Poems Children Will Sit Still For:*
"The Caterpillar," by Christina Rossetti
"Firefly," by Elizabeth Maddox Roberts
"Little Snail," by Hilda Conkling
"Mice," by Rose Fyleman

Fisher, Aileen, *When It Comes To Bugs*
"Spiders"
"Mrs. Brownish Beetle"

Milne, A. A., *Now We Are Six:*
"Forgiven" (An adult unexpectedly finds a child's pet beetle.)

Singer, Marilyn, *Turtle in July.*
"The Bullhead Fish"
"Dragonfly"
"Rattlesnake"

Fingerplays

Many of the traditional fingerplays are about the characteristics of small animals. Crouch down with the children to enact Vachel Lindsay's poem, "The Little Turtle."

The following is an animal shelter fingerplay.

Here Is a Bunny

Here is a bunny	(Fist with two fingers raised)
With ears so funny	
And here is his hole in the ground.	(Left hand on hip, arm curved)
When a noise he hears	(Stretch "ear" fingers)
He pricks up his ears	
And jumps in his hole in the ground.	(Thrust fist through curved arm)

—Author unknown

Add to this fingerplay to include the movement of small animals observed by children in your classroom.

Small Animal Parade

A slow, slow snail Drags down the trail.	(Stretch and contract right hand, dragging along left arm)
A looping earthworm Moves along with a squirm.	(Loop and wiggle one finger)
A spider runs past With eight legs so fast.	(Hands one on top of the other, tuck thumbs under, wiggle eight fingers)
The grasshopper springs With six legs and wings	(One hand crossed on top of other, thumbs and little fingers held under. Leap up and down)
The green and white frog Leaps over a log.	(Right fist leaps over left arm)
A seven-fin fish Swims by with a swish.	(Palms together, hands twist and turn, moving forward)

Art Activities

Easel Painting. One of the early shapes many children paint spontaneously is a loop with many strokes radiating from it. Often these are called spiders or bugs by the painter. Provide green and brown paint on insect-watching day and casually suggest that some children could have fun making

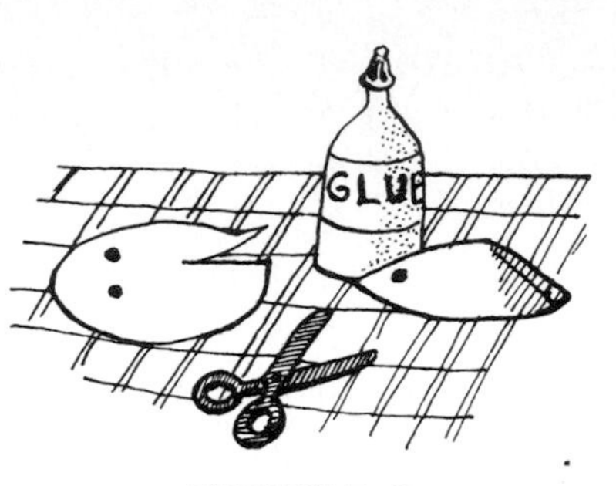

FIGURE 5–2

paintings of the spider, grasshopper, caterpillar, or whatever creature is of current interest. Newsprint sheets may also be cut into large butterflies or bird shapes for children to paint at the easel.

Paper Cutting, Shaping, and Pasting (Figure 5–2). To help children start this type of activity, have paper strips of $\frac{1}{2}$" by 4" (1 cm by 10 cm) and several sizes of oval shapes already cut for them. Show them that a strip can stand alone after being folded into a Z shape, or that it becomes a spring after being accordion-pleated. Suggest that such folds could become insect legs when pasted to body shapes. A single slash cut in an oval shape can be pasted over into a dart to create a three-dimensional shape from flat paper. Hang the resulting creature by a thread from a tree branch for an appealing mobile decoration.

Crayon and Picture Collage. Provide pictures of animals (for children to cut out or have them cut out already for those new at cutting), paste, construction paper, and crayons for crayon-enhanced collages. Animal pictures are not easy to locate in the typical household magazines. Old children's magazines, free nature publications from your state natural resources department, or animal stamps from the National Wildlife Federation are good sources.

Leather and Feather Collage. Scraps of leather from a crafts shop and assorted feathers make interesting collage material. Use durable paper and white glue for best adhesion. (Feathers might be unacceptable collage materials in classes where allergic reactions to feathers are a problem.) Feathers can be collected from poultry farms, or on late-summer walks near lakes or wooded areas. Feathers can be washed before use in this way: (a) Soak in cold water several hours; (b) Wash in warm water and detergent, swishing and surging water through the feathers; (c) Spread single layers on newspapers near a sunny window (no breeze, please). Let the children smooth the dry feathers with their fingers.

Additive Sculpture Animals. Pinecones make good bases for making animal forms. Turkeys can have rounded Scotch pinecone bodies, red painted balsam cone heads, and feather tails.

Use strips of molded paper egg cartons as caterpillar bodies. Offer tempera paint, fabric, paste, feathers, and snips of pipe cleaners for fanciful decorations.

Eggshell Mosaics. Broken, dyed eggshells can be glued to paper for egg-hatching project extensions. The younger the age group, the larger the shell bits should be for best management of materials.

Dramatic Play

Fishing. Add a dishpan full of water to indoor camping play. Tie a small magnet to a string. Attach to a stick. Cut simple fish shapes from pressed foam packaging sheets; attach a safety pin through it.

Animal Puppet Play. Put out animal hand puppets for children to animate. A small table turned on its side makes a good improvised stage.

Dramatize Animal Stories. When a group of children is familiar with an animal story that involves "a cast of thousands," they enjoy acting out the story in an informal, simplified way. Assign roles to all the children, so that no one is left out, even if someone has to play the part of a tree in the forest. Good animal stories that meet the casting requirements are: *The Story of Ping, Timothy Turtle, Make Way For Ducklings, The Tortoise and The Hare,* and *Caps for Sale.* Help move the story along informally, as needed. "One day Timothy Turtle was sliding down his favorite mud bank and he slipped and landed on his back . . . (Maria, you are pretending to be Timothy; can you roll over on your back now? Good! You're waving your arms and legs just as he did!)" Assign areas in the room for different scenes. "Here's the island on the river where the Mallards build their nest, and the block area is the Public Garden where they swim." Use simple props if they help carry out the story. For example, arrange a few blocks in a boat outline as Ping's houseboat, for the duck relatives to crouch in.

Food Experiences

Children accept the idea that people eat fish and chicken, since the terms are customarily used at mealtimes. There seems little to be gained, however, from pressing the point that the hamburgers being served for lunch were once steer or that the ham in the casserole was once a pig. Concentrate instead on ideas that foster positive emotions: cows giving milk, hens giving eggs, or bees making honey for people to enjoy.

To confirm human dependence on other forms of animals for food, many cooking experiences are possible in the classroom: making butter (use fresh, not sterilized, cream), instant puddings, ice cream, custard, and eggnog.

Thinking Games: Logical Thinking and Imagining

Habitat Classification. Pin pictures of a tree, lake, and grassy meadow to a flannel board. Mount small pictures, or cut out simple felt shapes, of an insect, fish, frog, turtle, snake, worm, spider, bird, and others. Let the children tell which picture "home" the creature belongs near. Can some animals live in more than one kind of "home"?

I'm Thinking of an Animal. Offer simple descriptions for children to guess, like: "I'm thinking of an animal that has no eyes, no legs, no arms; it has a mouth and a long body. What could it be?" Children who are experienced enough to take turns guessing will enjoy describing small mounted pictures drawn from a bag.

What If? "What if we had six legs to travel on? How would our lives be different? What could we do that we can't do with two legs? What couldn't we do? What if we had to spend most of our day gathering food for ourselves? What if we had working wings? How would our lives be different? What if we had to build places to live with no tools?" Encourage as many imaginative responses as possible. Let children generate their own "what if" questions to explore.

Animal Inventions. Encourage children to invent new animals, name them, draw what they would look like, and describe them.

See Joe Kaufman's *Big Book of Mammals and Birds* for more creative thinking ideas.

Field Trips

The list of potential problems and hazards that might be part of an outdoor animal observation hike could rule out this kind of field trip for a large group of young children. It can also be very disappointing when the objects of the trip do not display themselves. Try instead to pause for watching time whenever you encounter small animals while outdoors with your children: ants at work in a sidewalk crack nest, limp worms washed out of their tunnels after a hard rain, or squirrels and birds gathering food.

Other field trip possibilities are to a farm, zoo, pet shop, or animal shelter. The state conservation department will have information about the location of animal preserves, hatcheries, or parks with programs by naturalists. They may also have free or inexpensive leafiets about animals.

Do not forget the grocery store as an animal study resource. Animals provide, indirectly, our many dairy products, eggs, and honey.

Creative Movement

It's easy to draw children into the fun of acting out animal characteristics. Creative movement suggestions may also strengthen children's recall of the life cycles of animals that undergo metamorphosis. Slowly and dramatically read animal poems aloud for children to interpret with you. Good poems to try include the following found in *Poems To Grow On,* by Jean McKee Thompson (see Appendix 1):

"Cat," by Mary Britton Miller
"Mrs. Peck-Pigeon," by Eleanor Farjeon
"Baby Chick," by Aileen Fisher
"Fuzzy Wuzzy, Creepy, Crawly" by Lillian Schulz
"Earth Worm," by Mary McBurney Green

Clare Cherry suggests many animal-movement stimuli in her book, *Creative Movement for the Developing Child.* The Nature Company offers an imaginative sing-and-dance-along videotape, *Baboons, Butterflies & Me,* blending African wildlife footage with children imitating those movements. For 2- to 6-year-olds (see Appendix 4).

For an organized indoor play activity, children can be invited, one by one, to travel across the floor like the animal of his or her choice. (Groups of young children wait turns more patiently if they are sitting on the floor before and after their turn to travel.) One quick-thinking pair of 5-year-olds brightened our day when they piggybacked to crawl together as an 8-legged spider.

SECONDARY INTEGRATION

Keeping Concepts Alive

A teacher vividly reinforces concepts about the usefulness of insects by reacting calmly to an intruding bee or wasp. Remind children that insects of this sort sting only when disturbed. Open the windows from the top so that the insect can eventually fly out. Offer an observation jar to a shrieking child who is about to squash an uninvited spider. Do not undo what you have previously taught about the creature's place in the web of life by joining the chase with a can of insecticide. There is no need to be solicitous toward roaches, flies, or mosquitoes. If you must eliminate them in front of children, mention that they are pests to people but good food for birds and toads.

Another reinforcement of respect for living creatures may occur if a classroom pet dies or a dead bird is found on the playground. Let the children be aware of the death and take part in a gentle burial. Say that the animal's useful life has ended. Try to read one of the stories about the death of animals that coincides with your views of death (see Resources About the Death of Pets earlier in this chapter).

Relating New Concepts to Existing Concepts

The broad ecological relationships between plants, animals, soil, water, air, and temperature are awesome and complex. Preschool and early elementary-age children can make beginning steps toward understanding ecological relationships through small, concrete instances. For example, a child may observe a bird pecking at tree bark and wonder aloud if the bird is hurting the tree. An adult can clarify the bird's immediate purpose, and expand the idea of mutual dependency. "Yes, it does look a bit like the bird is hurting the tree. Really, it is getting its food, and helping the tree at the same time. The bird is catching tiny insects that are eating the tree. Can you think of a way that the tree can help the bird?" Emphasize that plants and animals need each other.

A child may question the fitness of animals' eating other animals, plants, and seeds for sustenance. One can appreciate the child's concern. Then suggest that each living thing can make many more seeds or eggs than are needed to make a young plant or animal like itself. If all the seeds grew into plants and all the eggs became animals, there would not be enough room in the world for all the living things. Some of the plants and animals need to be used in this way to keep the proper amount of each kind growing well to maintain the balance of nature. Young children can be upset by the recognition of our human role in the food chain. The fact that humans are animals that eat other animals may be better tolerated after the primary-grade years.

Sound concepts can be linked to the study of animal life. Compare the high pitch of the wren's song with the low pitch of the dove's coo. Feel the vibrations of a purring cat. Think of the swelling air pockets that make a frog's loud croak possible. Recall the vibrating wing covers that can be seen when the male cricket sings. Relate air concepts to the life requirements of animals (see chapter 7, What Is Air?).

Family Involvement

Inform families early in the school year that children are encouraged to bring in captured insects. After the first few planned experiences you may never have to track down a specimen by yourself.

A knowledgeable parent might be willing to set up a classroom aquarium, make a pet cage, or arrange to bring in and show a family pet. His child's social standing with the group usually blossoms as a secondary benefit.

EXPLORING AT HOME*

Watch and Wonder. Pause for a moment to watch the commonplace small animals you see when you are outdoors with your child: ants, worms, pigeons, squirrels exist nearly everywhere, even near city sidewalks. When you share that time together, you are teaching your child to respect our wild-animal neighbors. With some coaching, even a lively toddler can learn to be still and watch.

Children can only observe at home the insects, birds, and other small animals that come out at night. Some can be seen or heard in the city. Others can be seen or heard in the country or woods. A nature center near you might offer night hikes to observe signs of those animals. This is a special adventure that school cannot provide your child.

Try attracting birds to your yard or windowsill with a simple feeding arrangement. To learn more about how to do this read: the Golden Guide Book *Birds*, by Ira Gabrielson and Herbert Zim, or *The Bird Feeder Book*, by Donald and Lillian Stokes.

*These suggestions may be duplicated and sent home to families. Emphasize that these shared activities enhance what the child is learning, but they are meant to be enjoyed together. They are not homework assignments.

RESOURCES

ECHOLS, JEAN. *Animal Defenses.* Berkeley, CA: Lawrence Hall of Science, 1990. Creative activity plans for exploring animal defenses.

______. *Buzzing a Hive.* Berkeley, CA: Lawrence Hall of Science, 1990. Art and drama activities for exploring behavior of bees.

ENVIRONMENTAL DISCOVERY UNITS. Inexpensive teaching guides for ecology study. National Wildlife Federation, 1412 16th St. N.W., Washington, DC 20036.

GABRIELSON, IRA, & ZIM, HERBERT. *Golden Guide to Birds.* New York: Golden Books, 1987. Pocket-size paperback.

LEVI, HERBERT & LORNA. *Spiders and Their Kin.* New York: Golden Books, 1990. Pocket-size paperback.

MCFARLANE, RUTH. *Making Your Own Nature Museum.* New York: Franklin Watts, 1989. Practical tips are offered for safe collecting, careful preserving, and displaying varied specimens.

MITCHELL, ROBERT, & ZIM, HERBERT. *Butterflies and Moths.* New York: Golden Books, 1977. Pocket-size paperback.

RUSSELL, HELEN. *Ten-Minute Field Trips: Using the School Grounds for Environmental Studies.* Chicago: J. G. Ferguson, 1973.

STOKES, DONALD & LILLIAN. *The Bird Feeder Book.* Boston: Little, Brown, 1987 Paperback.

TUFTS, CRAIG. *The Backyard Naturalist.* Washington, DC: National Wildlife Federation, 1988. Paperback.

WILKES, ANGELA. *My First Nature Book.* New York: Alfred A. Knopf. 1990. Simple bird feeding projects. Directions for housing caterpillars and hatching moths and butterflies.

6

The Human Body: Care and Nourishment

Young children yearn to grow and become strong. They are eager to know what is inside their bodies. The experiences suggested in this chapter contribute to clarifying misconceptions and overcoming worries that children have about their bodies. They help children take the beginning steps toward a lifelong responsibility for their own physical well-being. The following concepts are explored:

- Each person is unique
- We learn through our senses
- Bones help support our bodies
- Muscles keep us moving, living, and breathing
- We help ourselves stay healthy and grow strong
- Strong bodies need nourishing food

The goal of this chapter is to help children learn to value themselves as unique individuals, respect differences between individuals, and care for their bodies. Mere knowledge of health care rules may not lead children into healthy practices. Positive attitudes about health lead to positive actions. Such attitudes grow through identification with a trusted person or symbol. It is suggested, then, that a puppet with whom children can identify should introduce each activity. Specific experiences lead to ideas about individuality, the senses, bones, muscles, and the heart. Simple health care and nutrition concepts follow from these activities. A method for evaluating what children have learned about their bodies is included.

CONCEPT: **Each person is unique**

Introduction: At group time a simple hand puppet awaits animation deep inside a large grocery bag on the teacher's lap. The teacher sits quietly until the children are comfortable and focused upon the impending action. The dialogue between teacher and puppet could go something like this: "Benjamin, it's time to come out now to meet all the children I've been telling you about" (reaching into the bag to slip a hand into the puppet). A hesitant puppet peers over the top of the bag, swiftly looks from side to side, and disappears inside the bag, protesting, "Those children can't be the ones you talked about. You said each one was different, but all these children look alike to me." The astonished teacher replies, "How can you say these children are all alike?" The occupant of the bag replies, "They *are* all alike! They all have wonderful bodies." "That's right, Benjamin. They do have wonderful bodies that work in wonderful ways, but the children are different in the ways they can use their bodies." The doubtful puppet agrees to this, adding, "But they *are* all bigger than I am." "Oh, Benjamin, I think I understand now. Perhaps you didn't know what it would be like to be with so many children. Why don't you come to the top of the bag just to listen while each child tells you something that is special about herself or himself. Then you will feel more like joining us another day." The puppet complies while each child reports an identifying characteristic. The teacher should be ready to add one positive statement about what each child can do. If the group includes identical twins, careful distinction should be made between the two.

When all of the children have been presented, the puppet acknowledges that each child really is special. He suggests that the children find out more about their special selves and begin to make *All About Me* books.

1. All about me.

LEARNING OBJECTIVE: To gain understanding of bodily self-image.

MATERIALS:

Manila folder or construction-paper cover for each child

Paper punch, fasteners

Crayons

Paper

Yardstick or tape measure

Scale

Full-length mirror or several smaller mirrors

GETTING READY:

Prepare book covers; label *All About Me,* (child's name).

Label page 1, "I am _____ tall, I weigh _____ , (date)."

Label page 2, "I look like this."

Fasten tape measure to wall securely, if used.

SMALL GROUP ACTIVITY:

1. Measure and weigh each child. Let child record the data on page 1. (Teasing about size differences hurts and must not be allowed.)
2. Encourage children to tell what they see in the mirror. Then draw what they see on page 2. Record each child's self-description on the page. (A withdrawn child may be unable to describe or draw himself. The teacher can help build the child's self-confidence by describing what the child's face looks like to her.)
3. Fasten completed pages in covers and tack to bulletin board until the next book entry is made.

Read *Is It Hard, Is It Easy?* by May McBurney Green.
Play "Everybody's Fancy," sung by Fred Rogers on *Won't You Be My Neighbor?*

2. *Mine alone.*

LEARNING OBJECTIVE: To appreciate one's uniqueness.

MATERIALS:

Two *All About Me* pages

Extra paper

Marker or crayon

Soft paper towels

Meat trays

Dry or mixed tempera paint, dark color

Water

Magnifying glasses

Scissors

Paste

GETTING READY:

Make paint pads for finger-printing by folding damp towels to fit trays. Sprinkle dry tempera or pour thick mixed tempera onto damp pad.

Mark book pages "Mine Alone, Right" and "Mine Alone, Left." (Children's names should be put on pages promptly to avoid confusion.)

Handprints of the class members make a meaningful bulletin board display for a parent meeting, stressing that individuality is to be celebrated.

SMALL GROUP ACTIVITY:

1. Invite children to trace around their hands, fingers spread wide, with marker or crayon, using appropriately marked pages. (Offer help if needed.)
2. Compare hand tracings for similarities and differences. Suggest looking at palms and fingertips through magnifying glass. How do they look?
3. Show children how to lightly stamp fingertips on paint pads, then press firmly on extra paper. Encourage examining the prints with magnifiers to compare patterns of each fingertip. Discuss the uniqueness of prints and the use of finger- or footprints when babies are born in hospitals. Mention that children are printing patterns that are theirs alone.
4. After children have enjoyed making many prints, suggest making a careful print in the corresponding fingers of the hand tracings. Children with mature cutting skills may wish to cut out the completed hands to paste in their books.

CONCEPT: We learn through our senses

Appreciation for the perceptual functions of our senses can be vividly experienced through their temporary absence. The following experiences make use of this principle:

1. *We learn by hearing.*

LEARNING OBJECTIVE: To appreciate hearing and to develop an understanding of hearing impaired persons.

Numbers are significant when they tell a boy how much he has grown.

Group Activity: Introduce this experience by carrying on a soundless, but lively, pantomime between teacher and puppet. Ask the children how they felt about the action. Could they tell what you and Benjamin were sharing? What was missing?

Demonstrate what story time would be like for a child who could not hear by reading silently while holding up a book to share the pictures. If a television set is available, watch a program with the sound turned off. Encourage discussion about what it would be like to have a hearing loss. Would it be hard to learn and to play with friends at school?

Read *A Button in Her Ear,* by Ada Litchfield. This story is especially meaningful if there is a hearing impaired child in the class. Without specific guidance, children find it hard to empathize with the limitations of a classmate whose hearing loss has no external signs.

Discuss how our ears function (see chapter 14, Sound). Discuss ear safety hazards. Try to arrange a visit from the public health nurses who conduct audiology screenings or from a hearing impaired person who is willing to show children his hearing aid or demonstrate sign language. Talk about how much we learn about the world through our hearing. Take a listening walk to encourage children to appreciate their ears.

2. *We learn by seeing.*

LEARNING OBJECTIVE: To appreciate vision and to develop an understanding of visually impaired persons.

Introduction: From his grocery bag Benjamin can be heard calling, "Help, I can't see. Where am I? Please help me." When the teacher helps him to the top of his bag, his problem is evident. A large paper cup covers his head. He explains that he was trying to get the last drops of milk from the cup when it slipped and got stuck on his head. After the teacher removes the cup, Benjamin comments gratefully that now he can see the beautiful faces around him. He would have had a hard time knowing who was here or what was happening when his eyes were covered. He asks the children if they could tell what was happening to him in the bag. If they couldn't see him, how did they know that he was in trouble? They got some information, but could they understand just what was happening when they couldn't see him? The teacher invites the children to find out what it is like to know things and to move about without help from their eyes.

MATERIALS:

Lengths of soft paper toweling, long enough to fold into thick blindfolds for each child (To avoid spreading eye infections do not share blindfolds)

Paper clips to secure blindfolds

Small objects to identify by touch. Mix easily identified items (brush, feather) with harder items (penny, nickel, 2 different books)

Small mirrors

Collection of objects to identify by sound: bell, clock, pieces of sandpaper to rub together, rubber band to snap

SMALL GROUP ACTIVITY:

1. Fasten blindfolds. Since it is hard for children to stay blindfolded for more than a few minutes, pass only a few items from child to child for them to identify.
2. Loosen blindfolds. What did the children find out?
3. Fasten blindfolds. Encourage children to identify sounds. Remove blindfold to let children use both eyes and ears to identify the sounds.
4. Let children examine their eyes in the mirrors and report what they see. Are the eyes still or moving? What do the lids do? How do their eyes feel when they gently hold lids open without blinking for a minute? Is someone in the group wearing glasses? Why?
5. Offer to lead a blindfolded walk in the room. Fasten blindfolds for those willing to participate. Put children's hands on shoulders of the child ahead. Slowly walk through the room, stopping to ask children where they are and how they know.

Read *Look At Your Eyes,* by Paul Showers and *A Cane in Her Hand,* by Ada Litchfield. Children could gain empathy for the visually impaired by trying cane traveling.

3. We learn by smelling and tasting.

LEARNING OBJECTIVE: To recognize that some senses work together.

Introduction. Benjamin greets his friends with a tissue in his hands. He uses the tissue as he sneezes, commenting that he helps his friends stay well when he covers his sneeze this way. He explains his funny speech saying, "By dose does dot work well today." Benjamin complains about two other annoyances: his breakfast lacked flavor today and he can't smell anything. The teacher consoles Benjamin, then suggests that the children could find out more about tasting and smelling at the science table.

MATERIALS:

Apple

Potato

Two bowls

Water

Lemon juice

Paring knife

Salt

Sugar

Gelatin dessert granules

Paper towel for each child

SMALL GROUP ACTIVITY:

1. Recall Benjamin's problem. Ask children to hold their noses closed. "Let's use our tongues to find out what we are eating. Taste these slices and decide what they are." (If the two foods look very different, have children take turns to help each other taste with eyes *and* nostrils closed.)
2. "Now taste your slices without holding your nose. How do they taste now?" Encourage more experimentation with tiny amounts of salt, sugar, and gelatin granules to find out if tongues alone tell us how food tastes.

GETTING READY:

Peel apple and potato. Slice thinly and drop into separate bowls of water. Add a few drops of lemon juice to prevent darkening of the slices.

Read *Follow Your Nose,* by Paul Showers.

4. What does our skin tell us?

Introduction: Benjamin appears briefly to ask the children to find out one thing about the contents of two containers on the science table without lifting or opening them.

MATERIALS:

Ice cubes or very cold water

Hot water

Two small containers with tight lids

Feathers

Chalk

GETTING READY:

Fill one container with ice water and the other with hot water. Cover with layers of paper toweling to retain temperature differences until children can use them.

SMALL GROUP ACTIVITY:

1. "What can you learn about what is inside the containers without opening them? How can you find out?" Which part of their bodies did they use? Could they find out the same thing by using their elbows? Foreheads? Backs of their hands? What covers each of these body parts?
2. Suggest finding out if it is easy for our skin to inform us about what it touches. "Try stroking your skin so gently with a feather that you can't feel it. Is it possible?
3. Ask whether skin on some parts of the body gives better information about how things feel. "Does chalk feel the same when stroked across an arm; when rubbed by the fingertips?"
4. Encourage children to bring in an object with a special texture to put on a "Please Touch" tray.

The following four experiences deal with anatomy. One way to evaluate whether children develop more accurate concepts about their bodies involves tracing their body outlines on sheets of newsprint before and after completing these activities. Both times ask the children to complete the picture by drawing the parts inside the body. Record each child's description of what has been drawn *as the child conceives it.* (Suggested by Dr. Robert Fitzmaurice, Science Educator, University of Houston.)

CONCEPT: Bones help support our bodies

LEARNING OBJECTIVE: To recognize that bone structure and body actions are related.

Introduction: Benjamin wants to share his thoughts with the children today. He has thought of another way that all of the children are alike. When they sit down they don't go limp and slide into a heap. When Benjamin isn't with the children, he flops to the bottom of his bag. He would like to know what holds them up. Can they tell him? Can they feel something inside their arms, legs, hands, and under their hair? Can they feel something in the middle of another child's back? He encourages them to investigate. Perhaps the children could find out more about bones at the science table and report their new ideas to Benjamin.

1. We have long bones and short bones.

MATERIALS:

Chicken rib section (mealtime leftover)

GETTING READY:

Remove bits of breast meat from chicken rib section, leaving connective tissue and ribs intact.

Assemble "Mr. Bones."

SMALL GROUP ACTIVITY:

1. Examine chicken rib section. Point out the connective tissue joining ribs. Help children feel their ribs, tracing the curve of ribs from front to back. Are children's bones harder? Longer? (Bring out the idea that children grow bigger as their bones grow longer.)
2. Encourage children to curve their backs; to curl up tight. Feel small bones in the spine; compare with long bones in arms and legs. "Why do arms and legs bend differently than backs?" Feel joints that connect the long bones in arms and legs.
3. Show a completed "Mr. Bones." Clarify that straws only suggest how long bones and backbones are different. (Each of our 206 bones has a special shape, unlike straw pieces.)

Read portions of *Your Living Bones,* by Louise McNamara and Ada Litchfield.

To assemble "Mr. Bones": Don't be discouraged by the lengthy directions. Once the straws and wire are cut, it is easy to create a figure from ordinary, inexpensive materials.

MATERIALS:

Large needle

17" (43 cm) and 30" (75 cm) lengths light wire (wires from telephone cable work well)

Utility scissors

Plastic drinking straws

4 twist closures

Cellophane tape

3" (7 cm) square light cardboard

Cotton ball

ASSEMBLY:

1. Head: Gently fold two wires in half. Slip fold over two fingertips; twist both wires together forming an oval loop. (Use $3\frac{1}{2}$" (9 cm) of wire in loop.) Thread wires through (e) size straw for neck. (Figure 6–1, step 1.)
2. Thread two shortest wires through top hole pierced in shoulder (a), feeding one wire out through straw to right, one out to left, forming arm bases. Thread two longest wires straight through both holes pierced in shoulder, forming trunk base. (Figure 6–1, step 2.)
3. For each arm, thread wire through straws (b) and (d). Bend remaining ends of wire into small loops for hands, twisting wire at wrists. Tuck ends of wires back into lower arm straws (d). (Figure 6–1, step 3.)

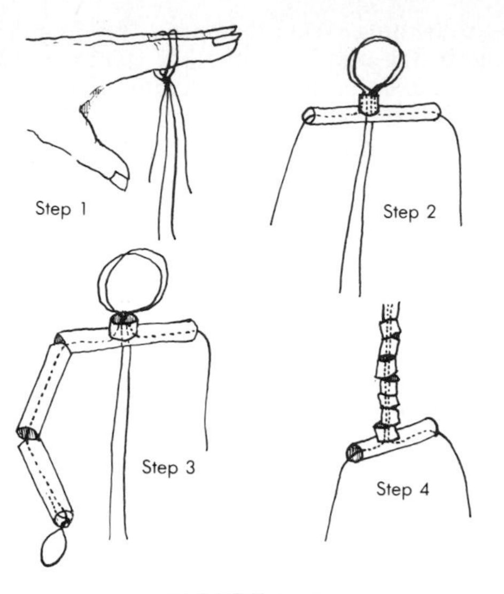

FIGURE 6–1

GETTING READY:

Cut straws into these lengths:

(a) $1\frac{3}{4}$″ (4.5 cm) 1 shoulder

(b) $1\frac{1}{2}$″ (4 cm) 2 upper arms and 2 thighs

(c) $1\frac{1}{4}$″ (3 cm) 1 pelvis and 2 lower legs

(d) 1″ (2.5 cm) 2 forearms

(e) $\frac{1}{4}$″ (1 cm) 13 vertebrae

Pierce center of shoulder straw piece with needle, making holes through top and bottom large enough for wires to pass through.

Pierce center of pelvis straw piece with needle, making only one hole through top for wires to pass through.

4. Thread trunk wires through 12 vertebra pieces (e), holding framework upside down to keep pieces in place. (Figure 6–1, step 4.)
5. Thread trunk wires through hole pierced in pelvis straw (c). Feed one wire out through straw to right, one out to left, forming leg bases.
6. For each leg, thread wire through straws (b) and (c). Bend 2″ (5 cm) of remaining wire into loops for feet; twist wire at ankles. Tuck ends of wires back into lower leg straws (c).
7. Bend feet at right angles to legs. Tape feet to cardboard, allowing Mr. Bones to stand.
8. Fasten 4 wire twist closures between upper vertebrae. Curve ends around to front, forming rib cage.
9. Stuff small amount of cotton into head loop. Fasten by winding tape around it, making head. Figure 6–2 shows completed "Mr. Bones."

CONCEPT: Muscles keep us moving, living, and breathing

1. Muscles keep us moving.

LEARNING OBJECTIVE: To identify the feeling of muscles in many parts of the human body.

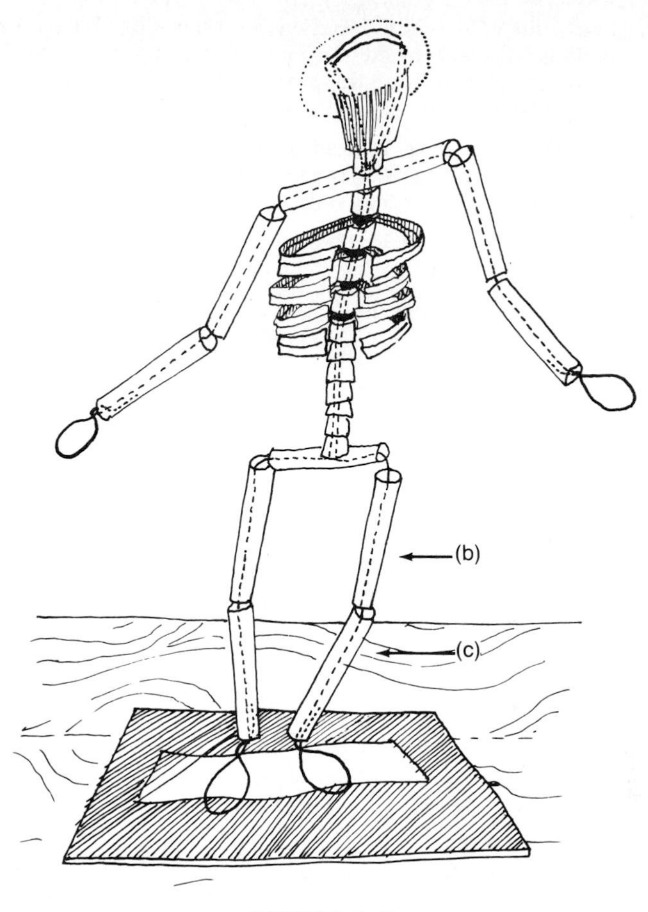

FIGURE 6–2

Introduction: Benjamin has persuaded the teacher to sew a drinking straw on each of his sleeves so he can pretend to have bones. He appears at the top of his bag, eager to show off his "bones," but something isn't quite right. "My arms stick out now, but that's all they do. They don't go down, they don't go up by themselves, and they don't wrap around you in a hug. You children can move your arms. What helps you move and stretch and lift?" Benjamin encourages the children to consider whether muscles exist only in the arms (a common misconception among children) or if they are found throughout the body. The discussion should bring out the fact that it is easy for many people to use their muscles without even thinking about them. Other people have to think and work hard to use their muscles. They may need to have help from special braces, crutches, or chairs.

The teacher can add that we have hundreds of muscles in our bodies. "Let's find out if we can feel some of them working to move our bodies."

Group Activity: This pattern for exploring the use and feeling of muscle groups is best done together, lying on the floor, but can be adapted to seated groups.

1. "Lift your fingers off the floor. Move just your fingers while the rest of your body is still and relaxed. Stretch your fingers high. Do you feel pulls in your hands and your wrists?"

2. "Now lift your whole hand off the floor. Lift just your hands while the rest of your body is still and relaxed. Turn your hands in circles. Do you feel pulls in your wrist and your arms?" Continue this way with lower arms and then with whole arms. Explore the pulls felt when the extended arm is slowly moved back and forth.
3. Continue the quiet, relaxed investigation of toe, foot and ankle movements, asking children what they feel as the body parts are slowly lifted, twisted and turned. Can they lift their lower legs only when lying on their backs? Why not?
4. Explore the muscles involved in curling up tightly, then slowly unfolding, sitting up, and twisting and turning the trunk.
5. Stretch the shoulder, neck, and jaw muscles. Appreciate smoothly working muscles. Conclude by using the facial muscles to make a smile.

Read: *You Can't Make a Move Without Your Muscles* by Paul Showers.

2. *Your heart: A powerful muscle.*

LEARNING OBJECTIVE: To identify the heart as a muscle.

Introduction: The teacher notices that Benjamin has been very quiet in his bag. Benjamin explains that he was trying to be perfectly still, not moving one muscle. It was hard to be so still. The teacher replies that some muscles *never* stop moving, even when we sleep. "We couldn't live if the mightiest muscle of all stopped moving; it is our blood-pumping muscle, the heart. The heart pushes blood through thin tubes to send energy and oxygen to all parts of the body. We can't see our heart working because it is safe inside our rib cage, but we can feel it working (demonstrate location)."

Show the children how to fold their hands together, then squeeze and release the grasp rhythmically. "Your heart works something like this, pumping and pushing day and night. Let's see how long we can do this with our hands. We can stop pushing our hands together when we get tired, but our hearts can never stop pumping."

MATERIALS:

Paper towel tubes

16" (40 cm) lengths of garden hose

Watch with second hand

Stethoscope (desirable)

SMALL GROUP ACTIVITY:

1. Ask children to fold their clasped hands next to their ears to listen as they squeeze their hands together. "Your heart makes a soft thumping sound like that each time it pumps and relaxes."
2. Children can hear each other's heartbeat using a paper tube, with one ear directly on the tube, the other ear covered with a hand. Children can hear their own hearts beating using the hose. (If a stethoscope is used, wipe earpieces with alcohol after each child's use.)
3. Try counting heartbeats for half a minute when children are quiet. Compare with heart rate after children jump in place for half a minute. What happens during exercise?

3. *What Pushes Our Lungs?*

LEARNING OBJECTIVE: To recognize that muscles keep us breathing.

Group Experience: Benjamin moves about vigorously inside his paper bag. He appears at the top of the bag panting heavily. He reports that he has been jumping hard and now two things are happening to him. He asks the children what they heard through the

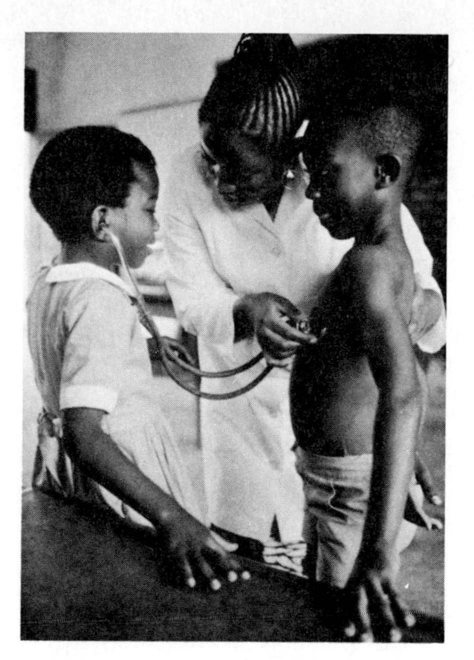

Robert's heart pumps, the visiting nurse facilitates, and Regina listens to life.

heart-listening tubes after they jumped a lot. Benjamin asks the teacher if it is all right for a real heart to work so hard. The teacher assures him that it is *more than* all right. "Heart muscles grow stronger just as other muscles grow stronger and work better when they are exercised well." Children may have much to contribute about muscle-building exercises.

If the children don't mention that they were out of breath after jumping in place, let Benjamin inquire about why he was panting after exercising. The teacher replies that hardworking muscles need a fast supply of blood and oxygen to get energy. The heart works hard to pump blood faster and other muscles speed up to pull oxygen from the air into the lungs. Children can find out more about how these muscles push the lungs in the following activity:

MATERIALS:

"Mr. Bones"

Yardstick (or meterstick)

Large mirror

Card or stiff paper for each child

SMALL GROUP ACTIVITY:

1. Look at the curved rib cage on "Mr. Bones." The rib cage protects important parts inside our bodies like the heart and lungs.
2. Have each child find a lower rib, then hold a card at right angles to the rib. What can be seen in the mirror? Is the card moving? How? Hold the yardstick next to a child as he or she watches the card rising and falling. How far does the card move? Special muscles near our midsections move our rib cages up and down. Why?

CONCEPT: We help ourselves stay healthy and grow strong

1. Our bodies need rest and exercise.

LEARNING OBJECTIVE: To recognize that rest is essential to ensure a healthy body.

Introduction: Benjamin has been exercising again. (If there are children in the class who are unable to use their legs, let the puppet "work out" with a weight made from a straw with two cardboard circles at the ends.) He doesn't understand why he feels tired. He has been running as fast as he can to make his heart and breathing muscles strong so he will feel really good. He learns from the teacher that after a hard workout muscles can use up the nourishment and oxygen brought by the blood. When that happens muscles lack energy to keep working, so they must relax and rest before doing more work. Our bodies get tired from working hard: Our eyes get sleepy, our minds don't think as well, or we may feel cross. Children have another important reason for slowing down to rest. *Their bodies need extra energy to grow.* When they rest and sleep at night (about 10 hours) their bodies can get ready for another day of working, playing, and growing. Benjamin suggests that children put pictures in their *All About Me* books to help them remember to exercise *and* rest.

MATERIALS:

Paste

Scissors

Crayons

Magazine or catalog pictures of active and resting people

GETTING READY:

Mark *All About Me* pages "I will give myself enough sleep each night" and "I will give myself enough exercise each day."

SMALL GROUP ACTIVITY:

1. Encourage older children to draw themselves enjoying exercise. Let children cut and paste pictures that remind them to exercise and rest.
2. Help children complete a realistic statement about how much sleep they require to stay healthy.

2. Soap washes away dirt.

LEARNING OBJECTIVE: To observe the cleansing effect of soap and water.

Introduction: Benjamin is humming as the teacher helps him out of his bag. He is carrying a bar of soap and a washcloth. The teacher is surprised that Benjamin seems unusually cheerful about washing. Benjamin replies that he has been thinking about all the things he can do for himself. "If I were a baby someone else would have to wash me. I am proud that I can take care of myself like a grownup." However, Benjamin is puzzled about why children in the class wash their hands before they eat. He asks why they wash their hands with soap, even when their hands don't look dirty. (The teacher may have to supplement the children's responses to help Benjamin understand that the organisms, or *germs,* that cause illness are too small to see. The organisms may still be on the hands when children eat unless soap washes them off.) Children could experiment to find out more about washing with soap and water.

MATERIALS:

Old white fabric

Two dishpans

Bar of soap

Water

Paper towels

Two trays

Paper and paste

Scissors

Potting soil

Oil

GETTING READY:

Experiment with water, soil, and small amounts of oil to make thin solution.

Cut off $\frac{1}{3}$ of the fabric to save as control samples.

Dip remaining fabric into soil solution. Let dry and cut into pieces, allowing two for each child.

Collect pictures of soap and of people washing.

Mark *All About Me* pages "I wash myself with soap."

SMALL GROUP ACTIVITY:

1. Let children wash one sample of dirty fabric in a dishpan of water and a second sample of dirty fabric in a separate dishpan of water, scrubbing the latter with a bar of soap.
2. Line two trays with paper towels. Put the water-washed samples on tray 1 and the soap-and-water-washed samples on tray 2. Mark children's initials next to damp samples.
3. Compare dried fabric samples to see which one is cleaner. Paste samples on booklet page next to a sample of the control fabric for comparison. Label. This may be a good time to provide an animal-shaped bar of soap to add fun to hand-scrubbing routines.

3. Immunization helps us stay well.

LEARNING OBJECTIVE: To develop acceptance of the need for immunization.

Group Experience: Benjamin has a book under his arm. He explains that he is going to be immunized today. "I'm going to have a shot. I don't want to have a shot, but it will keep me from being very sick . . . and I really do want to stay well." He asks the children how they feel about shots. After the children express their feelings, Benjamin asks the teacher if she would read a special book to help him get ready to go to the doctor's office. (*Mister Rogers Talks About Going To The Doctor* is a good choice for young children.)

Supplement the story with an opportunity for children to examine a disposable syringe *with the needle removed.* (A public health clinic might supply you with used syringes and directions for sterilizing them before allowing children to touch them.) Set up a hospital play situation. Invite a public health nurse to visit to provide information about immunization in a nonthreatening way.

Add a child-oriented message on immunization to the *All About Me* booklet (such as "Stay Well Cards" available from health foundations). Immunization record booklets from the public health office in your community could be sent home to parents to supplement the classroom focus.

4. *We should take care of our teeth.*

LEARNING OBJECTIVE: To recognize responsibility for one's own dental health.

Introduction: (Try to invite a dental hygienist to present this introductory information to the children. Some professionals are prepared to demonstrate good toothbrushing techniques with sets of plastic teeth and giant toothbrushes. Benjamin can stand in for the professional, if necessary.) Benjamin emerges from his bag with a toothbrush in his arms. He has just realized that he will never have a chance to use it since he has no teeth. All of the children already have 20 teeth that are as hard as their bones. They also have permanent teeth inside their jaws that grow bigger and bigger until they are ready to replace the baby teeth. Benjamin has no teeth, but he has a wish. He wishes that he could brush the teacher's teeth. The teacher agrees to the toothbrushing help, if Ben will brush each area round and round, then finish with downward strokes for upper teeth and upward strokes for lower teeth, the way the dentist suggests. Ben does a dry run, then asks the children if the job is done. The teacher supplements their suggestions *only* if they forget to mention that toothpaste helps to scrub away plaque if used twice a day. Flossing could be demonstrated for older children who already have permanent teeth. The teacher could add that people do two more important things for their teeth: They visit the dentist twice each year and they eat the foods that make strong teeth and gums, especially milk. They eat healthy snacks like fruit and nuts. Benjamin finds cotton-tipped swabs in his bag for each child to use to practice with him. He also finds a funny story about teeth, *How Many Teeth,* by Paul Showers, for the teacher to read.

5. *What do our teeth do?*

MATERIALS:

Small mirrors

Tissues

Graham crackers

Slices of apple

Magazine pictures of toothpaste, toothbrushes, and children brushing

Booklet pages

Paste

Scissors

Plaque disclosure kits (desirable)

GETTING READY:

Mark booklet pages "I Brush My Teeth."

Fold a tissue into a small pad for each child.

SMALL GROUP ACTIVITY:

1. Encourage children to smile at themselves in the mirror. What does the smile look like? Are teeth useful for more than smiles? Are teeth all alike?
2. Let children carefully bite down on the tissue pads with all their teeth. Look at the marks on the pads. Are they all alike? Which teeth would be best for biting? Which teeth would be best for chewing and grinding food fine enough to swallow?
3. Encourage using mirrors to examine the grinding teeth. Give children bits of cracker to chew. Use the mirrors to view the grinding surfaces again. Can the tongue clean away some cracker crumbs?
4. Offer apple slices to the children to chew thoroughly. Check with mirrors again. Has the apple, "nature's toothbrush," changed the way the grinding teeth look?
5. Join the children at the sink to practice swishing water around in the mouth to rinse away food bits.

If toothbrushing is part of the classroom routine, a few children at a time could try the plaque disclosure tablets after they have brushed their teeth. Adult supervision is needed since the disclosing agent stains skin, clothing, and surroundings. Provide the kits to parents if school settings aren't favorable for this activity.

CONCEPT: Strong bodies need nourishing food

1. *Which foods help us grow? Which do not?*

LEARNING OBJECTIVE: To distinguish between nutritious and nonnutritious foods.

Group Experience: Benjamin's grocery bag is flanked by a bag marked *P* and another bag marked *S*. Benjamin reports his experience visiting a grocery store. There he saw two elves buying food: Strong, who looked perky, healthy, and strong; and Pudge, who looked dreary and was almost as round as he was tall. Benjamin has pictures of the kinds of things each elf put in his grocery bag. Strong chose foods that help children grow well and have energy and strength. Pudge chose foods that make children grow pudgier and feel tired. Pudge's bag contains pictures of empty-calorie foods, especially those promoted in television advertising such as sweetened breakfast foods, candy, and snack chips. Strong's bag has pictures of nutritious foods and wholesome snacks. After the children have identified the pictures as they come from each bag, Benjamin gives a picture to each child. The child puts the picture back in Strong's *S* bag or in Pudge's *P* bag. Benjamin comments that Pudge's foods "fill you up and puff you out, but they let you down in health and strength."

Benjamin asks to hear the story *Bread and Jam for Frances*, by Russell Hoban. (Read *The Hungry Caterpillar*, by Eric Carle, for younger children.) Afterward the children can talk about the "junk" foods and healthful foods mentioned in the story. Also, the children could help choose pictures for a bulletin board display of what they have classified as healthful foods and junk foods.

2. *Choosing foods for growth, health, and energy.**

Group Experience: Assemble the following materials: plastic food replicas; empty containers of basic foods; and four grocery bags marked with individual labels and corresponding pictures for the milk group, protein group, vegetable and fruit group, and bread and cereal group.

Benjamin struggles to the top of his bag with a half-pint milk carton in his arms. He has been trying to plan a lunch that he and his friends can prepare, but his information about menu planning is limited. "I know that jam sandwiches, potato chips, and Popsicles taste good and fill us up, but they don't help us grow and stay healthy. We won't have that kind of lunch. I remember that we need milk to help teeth and bones grow strong. But if we have *just* milk, that would be a no-munch-lunch." The teacher compliments Benjamin for what he already knows, adding that it takes food from four different groups, working together, to keep us healthy, strong, and growing. Benjamin and his friends can tell more about nourishing meals after they decide which foods make up those four important groups. Slowly describe the nature and function of the foods in each group. Let Benjamin find food replicas as examples of each food group. Encourage the children to add their ideas and suggestions freely. (In general, the milk group builds muscle, bones, and teeth and supplies energy; the protein group* builds muscle, makes healthy blood, and supplies energy; the bread and cereal group (whole grains) keeps our bodies working well and gives energy; the vegetable and fruit group helps us stay healthy and keeps our bodies working well.)

Let each child choose a food replica and place it in the appropriate food group bag. Benjamin offers support and encouragement for children's beginning efforts at food group classification, then asks for help in choosing a balanced lunch menu. Ideally, this experience leads to making plans for the class to prepare lunch in the classroom. This can be a long-term project that involves a trip to the grocery store, food preparation, and lunch service. Parents can be served a nutritious child-made snack for a party, complete with recipe handout sheets.

*Note: The "Basic Four" food groups concept is no longer considered an adequate representation of nutrition for optimal health. At press time, however, no agreement has been reached among professional groups on how to symbolize today's definition of best nutrition. Be alert to the need for adapting these experiences when the "Food Pyramid" issue has been resolved.

*Because of world food shortages and the need for economical nutrition, vegetable protein sources should be emphasized as much as meat protein. The idea of combining vegetable proteins in one meal to yield complete protein values at a low cost should be communicated to parents as well.

3. Four food groups for our plates.

MATERIALS:

Pictures of food

Paste

Clear adhesive tape

All About Me pages

Brown rice, lentils, sunflower seeds

Plexiglass photo cube (desirable)

GETTING READY:

Trace a large circle on the booklet pages to represent a plate.

SMALL GROUP ACTIVITY:

1. If a photo cube is available, let children help you choose a picture representing each food group for four sides of cube. Top and bottom sections could have labels such as "Eat Food from Each Group Every Day."
2. Let children choose pictures of favorite foods from each group to paste on the "plate." Have the picture cube available as a reference.
3. Record child's description of the chosen menu on the booklet page.

If introducing complementary vegetable proteins, make separate strips of rice, lentils, and seeds between two pieces of tape. Cut into 1" (2.5 cm) lengths. Combine strips of rice and lentils on half of the squares of paper; lentils and seeds on other squares of paper.

4. Sprouting a vitamin crop.

LEARNING OBJECTIVE: To experience growing and eating nutritious food.

Note: Seed sprouting is probably the least expensive nutrition activity for the classroom. A few cents worth of seeds will produce enough sprouts for an appealing snack for the whole class. The sprouts are rich in vitamins and minerals. The sprouting process can involve each child during the plant germination period.

FIGURE 6–3

MATERIALS:

Alfalfa seeds from a natural foods store

Smallmouthed quart jar

6-ounce foam cup

Thick needle

5″ (12.5 cm) square of porous fabric

Rubber band

Cellophane tape, paper

Saucer

GETTING READY:

Method I

Pierce the bottom and lower half of the cup generously with a needle to form drain holes. (Figure 6–3.)

If class is large, prepare two sprouting jars to allow more child participation and provide more sprouts.

SMALL GROUP ACTIVITY:

1. Discuss the sprouting project with the children on Friday. Tape a few dry alfalfa seeds to a piece of paper to compare day-by-day with the changing sprouts.
2. Let children measure 2 tablespoons of alfalfa seeds into the jar. Explain that you will add one cup of warm water to cover the seeds in the jar at your home on Sunday night.
3. On Monday morning demonstrate how to cover the jar top with the fabric fastened with a rubber band so the water can be carefully drained off. Fill the jar with cool water and let children gently swirl the seeds in the fresh water. Then carefully drain the water.
4. For Method I, remove fabric and insert drain cup into the mouth of the jar as far as it will fit. Invert the jar so the cup rim serves as a stand. Place on a saucer. Keep it away from direct sunlight.
5. Set up a rinsing/draining schedule two or three times each day so that each child has a chance to care for the sprouting seeds. As the sprouts begin to develop, the foam cup can be used instead of fabric for draining and rinsing the sprouts for Method I.
6. As tiny green leaves appear, the jar may be kept in sunlight to develop chlorophyll. Serve sprouts on Friday.

FIGURE 6–4

Method II

Follow the same procedures as for Method I, except use a 2-liter plastic beverage bottle for a sprouting container. Cut off the top quarter of the bottle and slide a nylon knee sock over the open end. Knot the top of the sock as shown in Figure 6–4. Rinse and drain sprouts through the nylon mesh. Hold by knotted end to drain.

PRIMARY INTEGRATING ACTIVITIES

Music (see Appendix 1 for Music Book and Recording Information)

1. Raffi's *Singable Songbook* offers "I Wonder If I'm Growing?" and "Brush Your Teeth" to extend human body study. "Popcorn" and "Peanutbutter Sandwich" extend nutrition concepts.
2. Mister Rogers has recorded songs that extend human body concepts: "You are Special" in the album, *Won't You Be My Neighbor?*; "Everybody's Fancy" and "Everything Grows Together," in the album *Let's Be Together Today.*
3. *The Inside Story,* an album written and sung by John Burstein, is an excellent resource. "Mr. Slim Goodbody" has songs about nutrition, toothbrushing, the heart, and exercise.
4. "Here We Go 'Round the Mulberry Bush" provides a good pattern for improvising a song about health care. Children can sing, "This is the way I take

care of myself . . . so I'll be strong and healthy." Add verses such as "This is the way I (brush my teeth, take my bath, wash my hands, go to sleep, eat good meals) . . . so I'll be strong and healthy."

5. Young children like to contribute to a simple song about foods they enjoy. Use the tune to Ella Jenkins' song "I Like the Animals in the Zoo," from her album, *Seasons for Singing*. Each verse is a favorite food suggested by a child, such as:

   Lisa likes scrambled eggs, yum, yum, yum!
   Lisa likes scrambled eggs in her tum.

6. Listen to "What Foods Should We Eat Every Day?" and "How Does a Cow Make Milk?" sung by Tom Glazer on the *Now We Know* record.

Math Experiences

1. Translate the evidence of growth from birth weight to present weight into concrete experience by stacking enough hardwood blocks on a bathroom scale to equal the two weights for each child. Children may want to enter the following information in their *All About Me* books: When I was a baby, I weighed as much as _____ blocks; now I weigh as much as _____ blocks. Each child can feel good about her or his present state of growth when it is compared with past growth.
2. Establish a relative measure of muscle strength by weighing the number of blocks a child can lift. (Avoid competitive emphasis by using questions like, "How much weight are your muscles ready to lift now?")
3. Use fruit and vegetable shapes for flannel board math activities in set formation, one-to-one correspondence, and patterning.
4. Allow time for children to compare measuring spoons and cup sizes when they are used in cooking experiences. Try to have enough measuring equipment on hand so that all the children can participate in measuring ingredients. "We need one cup of flour for the dough. Will we have enough if two children each put in one-half cup of flour? Let's find out."

Stories and Resources About Individuality

ARTHUR, CATHERINE. *My Sister's Silent World*. Chicago: Children's Press, 1979. Photographs and story of a deaf child on her seventh birthday convey important insights about this handicap.

CHURCH, VIVIAN. *Colors Around Me*. Chicago: Afro-Am Publishing, 1971. Variations in skin colors among black peoples are discussed and illustrated.

FASSLER, JEAN. *Howie Helps Himself*. Chicago: Albert Whitman, 1978. The daily frustrations and pleasures, and a triumphant moment, in the life of a physically handicapped child are portrayed in this story. The sensitive illustrations by Joe Lasker help to develop understanding about handicapping conditions.

GREEN, MAY MCBURNEY. *Is It Hard, Is It Easy?* Reading, MA: Addison-Wesley, 1960. Promotes acceptance of differing levels of physical skills among a group of children.

KEATS, EZRA JACK. *Apt. 3.* New York: Macmillan, 1986. Sam learns to appreciate how his new sightless friend learns about his world. Paperback.

LASKER, JOE. *He's My Brother.* Chicago: Albert Whitman, 1975. Portrays some painful experiences of a child with the "invisible handicap," a learning disability. Empathy is encouraged for the child with a problem that is hard for others to understand.

LITCHFIELD, ADA. *A Button in Her Ear.* Chicago: Albert Whitman, 1976. The story follows the detection and correction of a child's hearing loss. Angela's teacher helps build acceptance among classmates for her hearing aid and battery harness.

______. *A Cane in Her Hand.* Chicago: Albert Whitman, 1977. Val encounters difficulties when her vision problem develops. She learns how to use a cane "like a long arm" to help her adapt to her limitation and carry on with activities she can enjoy.

______. *Captain Hook, That's Me.* New York: Walker & Co., 1982. A girl born with one hand copes with her limitation. For primary-grade children.

STERLING, BARBARA. *I'm Not So Different.* New York: Golden Books, 1986. Kit's story is designed to win respect and acceptance from children whose legs do for them what Kit's wheelchair does for her.

Stories and Resources About the Body

ALIKI, *My Five Senses.* New York: Harper & Row, 1989. A charming child enjoys experiencing the world through sensory awareness. Preschool. Paperback.

ALLINSON, ELAINE. *Daniel's Question: A Cesarean Birth Story.* Monsey, NJ: Willow Tree Press, 1981. A helpful book to recommend when cesarean parents need help in answering children's questions about their birth.

BALESTRINO, PHILLIP. *The Skeleton Inside You.* New York: Harper & Row, 1989. Describes the supportive and protective functions of our bone structure. A good classroom reference. Paperback.*

BRANDT, KEITH. *The Five Senses.* Mahwah, NJ: Troll, 1985. A good narrative style presents accurate information about senses as part of the nervous system. Paperback.*

COLE, JOANNA. *The Magic Schoolbus: Inside the Human Body.* New York: Scholastic, 1989. This clever fantasy trip down the esophagus into the stomach, bloodstream and major organs shows how the body converts food to energy for body cells. Highly recommended to lure children into respecting their bodies! Paperback.*

HOBAN, LILLIAN. *Arthur's Loose Tooth.* New York: Harper & Row, 1985. Arthur copes with the experience of losing his first tooth.

KAUFMAN, JOE. *Joe Kaufman's Big Book About the Human Body.* New York: Golden Books, 1987. Excellent beginning information about the structures of the body, laced with the author's lighthearted illustrations. The accurate information provides valuable correction for the common misconception that our lungs are like balloons.

MCNAMARA, LOUISE, & LITCHFIELD, ADA. *Your Living Bones.* Boston: Little, Brown, 1973. Effective illustrations by Pat Porter help keep the information about bone functions, growth, and physical makeup understandable to children.

SABIN, FRANCENE. *Human Body.* Mahwah, NJ: Troll, 1985. A good reference on anatomy and physiology for any early childhood classroom. Simple, clear illustrations by Don Sibley. Paperback.*

SHEFFIELD, MARGARET. *Before You Were Born.* New York: Alfred A. Knopf, 1983. Simple answers to the questions young children frequently ask about their prenatal development. Tenderly illustrated by Sheila Bewley.

SHOWERS, PAUL. *Follow Your Nose.* New York: Thomas Y. Crowell, 1963. Let's Read-and-Find-Out. This book includes an easy experiment that demonstrates the function of the nose.

*Starred references, written at the young child's level of understanding, can help teachers with minimal backgrounds in science expand their knowledge base.

______. *How Many Teeth?* New York: Thomas Y. Crowell, 1965. An amusing example of how teeth help us speak is part of this description of deciduous and permanent teeth.

______. *Hear Your Heart.* New York: Thomas Y. Crowell, 1968. This description of the heart and its functions explains the need for resting our bodies.

______. *Use Your Brain.* New York: Thomas Y. Crowell, 1971. Explains why it doesn't hurt to cut the nails and hair, and offers a first message about the effects of drug abuse on the brain.

______. *You Can't Make a Move Without Your Muscles.* New York: Thomas Y. Crowell, 1982. Good information about muscles, large and small. Children will have a hard time sitting without moving those muscles as they listen to this book, so let them move!

______. *A Drop of Blood.* New York: Harper & Row, 1989. Information, poems, and amusing illustrations reassure children that a cut will stop bleeding. Paperback.

SMITH, KATHIE, & CRENSON, VICTORIA. *Hearing.* Mahwah, NJ: Troll, 1988. The interactions between our sensory organs and the brain are described at a level that is simple enough to help kindergarten/primary-grade children appreciate these marvelous processes. Cheerful illustrations supplement this excellent information. Paperback.* Other books in this series are: *Seeing, Smelling, Tasting, Thinking,* and *Touching.*

VAN CLEAVE, JANICE. *Biology for Every Kid.* New York: John Wiley & Sons, 1990. Experiments with lung capacity measurement are included.

Stories and Resources About Health Care

BAINS, RAE. *Health and Hygiene.* Mahwah, NJ: Troll, 1985. General health care principles are described: good nutrition, rest, exercise, and cleanliness. Paperback.*

HOBAN, RUSSELL. *Bedtime for Frances.* New York: Harper & Row, 1960. Did Frances and her family get enough rest during a confusing night?

JACOBSEN, KAREN. *Health.* Chicago: Children's Press, 1982. The child is shown having some responsibility for his health.

REY, M., & REY, H.A. *Curious George Goes to the Hospital.* New York: Houghton Mifflin, 1966. George's experiences illustrate the hazards of swallowing inedible substances and offer some information about hospital care.

ROCKWELL, ANNE & HARLOW. *Sick in Bed.* New York: Macmillan, 1982. A simple story about getting well, told for young preschoolers.

ROGERS, FRED. *Mister Rogers Talks About Going to the Doctor.* New York: Platt & Munk, 1972. The low-key calmly presented text and photographs prepare young children for visits to the doctor and dentist.

______. *Going to the Dentist.* New York: G. P. Putnam's Sons, 1989. A First Experience Series book that introduces preschoolers to the dentist and to dental hygiene. Paperback.

SCARRY, RICHARD. *Nicky Goes to the Doctor.* New York: Green Press, 1978. Paperback. A routine medical checkup, including a shot, goes well for a young rabbit.

SHOWERS, PAUL. *No Measles, No Mumps For Me.* New York: Thomas Y. Crowell, 1980. A much-needed explanation of how immunizations work. For primary grades, but adaptable for younger children.

Stories and Resources About Nutrition

BAINS, RAE. *Health and Hygiene.* Mahwah, NJ: Troll, 1985. Clever illustrations clarify the function of nutrients needed to sustain our bodies. Paperback.*

BERENSTAIN, STAN & JANE. *The Berenstain Bears and Too Much Junk Food.* New York: Random House, 1985. Presents a good level of information about the need for nutritious food to maintain body systems. Paperback.

CARLE, ERIC. *Pancakes, Pancakes!* New York: Alfred A. Knopf, 1975. Jack's desire for a breakfast pancake is delayed while he gathers each ingredient directly from wheat field, barn, and chicken yard.

HOBAN, RUSSELL. *Bread and Jam for Frances.* New York: Harper & Row, 1964. Frances goes on an empty-calories eating binge. The problem is resolved humorously and effectively.

LINDSAY, TRESKA. *When Batistine Made Bread.* New York: Macmillan, 1985. Describes breadmaking from harvesting grain to baking the loaf in times before machines made the task easier.

MILNE, A.A. *Winnie-The-Pooh.* New York: E.P. Dutton, 1962. "Pooh Goes Visiting." Pooh overeats a sweet snack at Rabbit's house and suffers an inconvenient consequence.

MINARIK, ELSE. *Little Bear.* New York: Harper & Row, 1957. In the story "Birthday Soup," Little Bear and his friends share birthday soupmaking for a party with a surprise. A good take-off point for a class cooking project.

SCARRY, RICHARD. *What Do People Do All Day?* New York: Random House, 1968. "Mother's Work is Never Done" features children helping mother prepare meals. "Where Bread Comes From" illustrates the processes involved in breadmaking from wheat harvesting to the baking ovens.

SOLOT, MARY LYNN. *100 Hamburgers: The Getting Thin Book.* New York: Lothrop, Lee & Shepard, 1972. A rueful look at a young boy's overweight problem: what he can eat and what he wishes he could eat.

STEVENS, CARLA. *How to Make Possum's Honey Bread.* New York: Seabury Press, 1976. Possum shows four friends how to make bread. Bread recipes are included. A good prelude to classroom baking.

Poems (Resources in Appendix 1)

From *Now We Are Six,* by A.A. Milne, read "Sneezles."
From *Egg Thoughts and Other Francis Songs,* by Russell Hoban, read "Egg Thoughts."

Fingerplays

Use this fingerplay to encourage handwashing:

> This right hand is a very fine hand.
> This left hand is its brother
> Together they wash with soap and water.
> One hand washes the other.
>
> —AUTHOR UNKNOWN

Incorporate other healthcare principles into the familiar fingerplay, "This little boy climbs into bed. . . . "

Art Activities

Paper Doll Chain. To recall that people are alike in many ways, fold and cut out paper doll chains. Encourage children to use markers or crayons to make each paper doll different. "Make each one look special, just as each of you is special."

Individuality Collage. Make a collage poster as a group project using pictures of persons of varying ages and races to emphasize the uniqueness of each person.

Fingerprint Art. Children's fingerprints can create all-over designs and patterns. Use a simple paint stamping pad for prints (see chapter 4, Plant Life). Suggest using crayons to add stems to print-petaled flowers, strings on fingerprint balloons, and so forth.

Clothespin Dolls. Plain wood clothespins form head, trunk, and legs of dolls; pipecleaners twisted around the "neck" form arms and loop hands. Features can be drawn on the heads with ballpoint pen. Scraps of fabric can be fastened to the body with rubber bands. Compare the flexible and rigid parts of the doll bodies with the bone structure of the human body.

"Foods We Need Each Day" Poster. Let older children make a poster or bulletin board showing the number of servings of each food group needed daily for good nutrition. Use four different colors of construction paper as background for the collage, arranged so that the finished sheets make the 2-3-4-4 pattern evident:

2 blue sheets for two protein servings
3 yellow sheets for three milk food servings
4 green sheets for four fruit and vegetable servings (include a source of vitamin A)
4 tan sheets for four cereal and grain servings

Dramatic Play

Hospital Play. Add paper towel tubes to the props provided for hospital play so that children can listen to heartbeats. Add discarded plastic syringes (sterilized, needles removed) to allow children to take the role of shot-giver.

Grocery Store. Compare the ways that children arrange the food replicas provided for a play store before and after the nutrition experiences.

Restaurant. Make picture menus from manila folders, using illustrations of wholesome foods marked with large-numeral prices. Provide restaurant workers with pads of paper, pencils, a toy cash register, trays, and food replicas (or pictures cut from cardboard food cartons).

Thinking Games: Logical Thinking and Imagining

Alike and Different Game. Ask one child to look around at the other children to find someone who is like her or him in *one* way. If Elena chooses Jenny because they are both girls, ask Jenny to stand next to Elena and find one way that she and Elena are different. Then Jenny has a turn to find someone who is like her in one way, and so on. End the game with a comment about no two people being just alike in every way. Mention the good thinking and choosing the children did.

Four Food Groups for Our Plates. (Suggested by Polly Warner, Racine, Wisconsin Montessori School, as a self-correcting exercise for children to carry out independently.) Mark miniature food models with adhesive signal dots, color coded to indicate membership in one of the four food groups. Mark small paper plates into four sections and color each section to correspond with the food group signal dot colors. Let children create balanced meals by choosing miniatures from each food group, and placing them on colored plate sections that match the dot colors. Children store materials in shoe boxes, using the lids as work trays for their balanced menu plates.

A Fantasy Sensory Walk. (Suggested by Hope Jordan.) Tape-record outdoor sounds: twigs breaking, squirrels scolding, rocks splashing as they fall in a puddle, waves breaking, etc. Collect as many objects as possible, such as pine sprigs and cones, rocks, and apple slices. Later tell a story of a walk in the woods to closed-eyed children, who imagine themselves on that walk with you. In a hushed tone question what you are experiencing: "What brushed against my arm? What kind of fruit is growing on that tree?" Pass around the pine sprig, slices of apple, etc., as they fit the story. Ask similar questions while replaying the sounds. Afterward, make a chart listing the senses across the top. On the side, list the events or objects children recall encountering in the fantasy walk (dew on the grass, crow, apple). Ask "How did we know it was a pine branch . . . a crow . . . an apple?" Let children place a check under the senses that told them what the event or object was, completing the chart.

Creative Movement

Try to find a marionette to demonstrate its stiff movements controlled by strings. Compare the differences between the carved figure and children's bodies. Encourage the children to dance to a record, using their bodies as though someone else controlled their movements with strings. End the dancing by saying that someone has "cut their strings."

Show the children a rag doll, whose body is so flexible that it can be bent and curled in many ways that the marionette and children could not move. Let the children dance like rag dolls without bones or muscle control. Change the movements to those of real people who have muscles to stretch.

Various animal actions are suggested for creative movement in *Joe Kaufman's Big Book About the Human Body.*

Field Trips

Public Health Clinic. If a public facility can be toured, try to have the visit serve the children in a nonthreatening way, such as having children measured and weighed. Be certain that the children are prepared for what to anticipate at the clinic.

A Visit From the Field. Your community emergency medical service squad may be willing to bring a field trip to your school door. Children feel more at ease if a health professional visits them in a familiar setting. A careful look at the emergency vehicle can be fascinating and reassuring. The squad members evoke the awe of heroism, rather than the shadow of anxiety that may be aroused in some children when they meet health care professionals.

SECONDARY INTEGRATION

Keeping Concepts Alive

Teachers who are committed to the principle of accepting responsibility for one's own health and nourishment find it easy to use teachable moments to promote this goal. When a respected teacher joins the children in routine hand washing and toothbrushing, children emulate the patterns readily. When giving first aid to a scraped-knee victim, reassure the child that parts of the flowing blood are already beginning to make repairs so that new skin can grow over the scrape. Encourage children to discuss feelings about medical treatment when they have it and express respect for the way they faced what had to be done to keep themselves healthy.

Teachers have an excellent opportunity to help children appreciate non-sweet foods when they have snacks or lunch with the class. This is a natural time to discuss food and to talk about the food groups represented in the menu. Use the simple rule that everyone (including adults) tries to taste each food being served, allowing (rather than requiring) children to eat as much as they can. Avoid using food as a punishment or as a bribe to force children to eat disliked food. This practice weakens trust between child and teacher and may fail to win compliance from the child as well.

Undue "preaching" of health care and nutrition messages to dictate home practices can become counterproductive. Young children are limited in their ability to change the habits or cultural patterns followed by their families. Therefore, pressing children to conform to healthful standards away from school may result in feelings of guilt or resentment. The common practice of requiring children to report what they have eaten at home each day can threaten feelings of acceptance and reduce self-esteem. Neither practice may result in the goal of translating health care knowledge into actual health care practices. Teachers contribute most to this goal when they offer accurate information to children and parents, model sensible eating and health care habits, encourage children to follow their lead, and support those who try.

Relating New Concepts to Existing Concepts

1. When children are learning about their sense of vision, think about how light passing through curved glass makes things look larger (see chapter 15, Light). "This is how glasses help some people see better."
2. When discussing hearing, mention the delicate part inside of each ear that vibrates to let us hear sounds (see chapter 14, Sound).

3. Identify the calcium that our bones and teeth get from milk and dark green vegetables as the same mineral that makes hard shells for some animals (see chapter 5, Animal Life) and compresses or combines with other minerals to form certain rocks (see chapter 10, Rocks and Minerals).
4. When children are learning about the muscles that keep the lungs expanding and contracting, mention that all living things need air to stay alive.
5. Relate muscle development and exercise to simple machines by providing an inexpensive rope and pulley exerciser in the classroom.
6. Recall seed sprouting experiences (see pages 52–53). Each seed contains enough nourishment to start a new plant. That same nourishment also helps all parts of our body grow. There would be enough food to feed everybody in the world if we all would eat more protein from plants and less from meat. (See *Diet For A Small Planet,* by Frances Lappe.)
7. Relate the need for clean hands to the need for sanitary food preparation with this comparison: Seal unrinsed wheat berries in one plastic sandwich bag, together with a damp paper towel. Seal thoroughly rinsed wheat berries in a similar bag. Leave flat, undisturbed for two or three days. Examine the seeds with a magnifier several days later for clear evidence of mold or bacterial growth.
8. Experiment with two pots of a fast-growing plant such as coleus. Provide only one plant with liquid fertilizer each month. The difference in growth in a few months' time will illustrate how proper nourishment helps living things grow. (Make it clear, however, that nourishing substances for plants can make humans very sick.)
9. For some classes it might be desirable to explore the concepts in this chapter in smaller clusters during the school year. This sequence could be followed: individuality, sensory perceptions, anatomy and physiology, health care responsibility, and nutrition. The *All About Me* booklet could be developed over this period of time to serve as a source of continuity for the related concepts.

Family Involvement

Two goals of this chapter, valuing one's self and assuming responsibility early in life for one's own health, cannot be achieved without parental involvement. While few parents intend to neglect their children's health, there has been a marked decrease in communicable disease immunization for young children. The number of malnourished children in our affluent country is also discouraging. These problems are being addressed by the American Pediatrics Association campaign to urge prevention of illness through responsible self-care, rather than through reliance upon costly treatment to cure illness. Teachers and families together can encourage responsible attitudes in children. The *All About Me* booklet can spur families' interest in promoting healthy development for their children.

EXPLORING AT HOME*

Measuring Up. Your child is eager to grow. Children mainly grow taller as their long bones grow longer. Let your child gauge his or her growth against your full-grown size by comparing leg bone length with yours. How much do their leg bones need to grow to match yours? Do the same for hands, fingers, and feet. Let your child try on your sweatshirt or jacket. Is there still growing to be done to reach your size? Talk about the milk she or he needs to drink to help those bones grow longer. Mention that the bones grow fastest during sleep when our bodies aren't moving much. (A good reason for getting enough rest.) If possible, make a growth record inside a closet by penciling a mark at your child's present height. Record the date and inches on a piece of tape beside the mark. Do this each birthday.

Mark That Growth. Some parts of mature bodies continue to grow. Remind your child that even though you have stopped growing taller, your hair and nails are always growing, and need cutting many times each year. Together, observe the slow growth of your fingernails. Mark a dot of ballpoint ink next to the cuticle on one fingernail. Preserve the dot with a coat of clear nailpolish, if you have some. Find out how long it takes for the mark to move away from the cuticle. Compare your nail growth with your child's. Mention that skin can continue to grow throughout our lives.

Check This Out. Join your child in checking some interesting body measurement comparisons. Your child should stand with arms and hands outstretched. Cut a piece of string the length of the span from fingertip to fingertip. Now compare that string length with your child's height. They are probably about the same. Compare these two measurements on yourself and other family members. Next, compare the length of your child's foot with the length of his or her forearm, from wrist to elbow. Are the measurements about the same? Let your child measure and compare your foot and forearm. They are probably about the same, also.

RESOURCES

The Human Body

ALLISON, LINDA. *Blood and Guts: A Working Guide to Your Own Insides.* Boston: Little, Brown, 1976. Accurate information about anatomy and physiology presented in a lively style. Experiments suggested for older children may be adapted for younger children.

COBB, VICKI. *Keeping Clean.* New York: J.B. Lippincott, 1989. What would happen if we didn't keep ourselves clean? Simple information and clever illustrations make the answer clear.

*These suggestions may be duplicated and sent home to families.

GREENE, MARTIN. *A Sigh of Relief: The First Aid Handbook for Childhood Emergencies.* New York: Bantam Books, 1977. Easy, clear information and illustrations of first aid steps and child safety precautions. A logical choice for every school's professional bookshelf.

MULDOON, KATHLEEN. *Princess Pooh.* Niles, IL: Albert Whitman, 1989. A girl learns about the limitations experienced by her wheelchair-confined sister.

PARKER, STEVE. *Skeleton.* New York: Alfred A. Knopf, 1988. Outstanding color photographs and art enhance the basic information about the structure and function of the human and animal skeletal systems. London Museum of Natural History Eyewitness Series.

Nutrition

BURNS, MARILYN. *Good For You.* Little, Brown, 1978. Hard facts with a light touch about the digestive system and good nutrition. Interesting experiments for primary grades.

GOODWIN, MARY, & POLLEN, GERRY. *Creative Food Experiences for Children.* Washington, DC: Center for Science in the Public Interest, 1980.

LANSKY, VICKI. *The Taming of the C.A.N.D.Y.* Monster.* Wayzata, MN: Meadowbrook Press, 1978. (*Constantly Advertised, Nutritionally Deficient Yummies) Offers recipes for intriguing snacks that children enjoy making and eating. A sun-dried fruit leather recipe is included.

LAPPE, FRANCES. *Diet for a Small Planet.* New York: Ballantine, 1985. This book clarifies how world food needs can be met by producing more vegetable than meat proteins. Interesting, low-cost recipes are included.

WANAMAKER, N., HEARN, K., & RICHARZ, S. *More than Graham Crackers.* Washington, DC: National Association Education of Young Children, 1979. Nutrition education and food preparation with children.

Additional Resources

NUTRITION CURRICULUM. The National Dairy Council offers a free nutrition education materials catalog. Branch offices across the country where films, materials, and curriculum kits can be borrowed are listed. Write to the National Dairy Council, 6300 River Road, Rosemont, IL 60018.

FAST FOODS EATING GUIDE. The Center for Science in the Public Interest offers a poster evaluating the nutritional values, or their absence, of favorite fast food restaurants' menu items. To order, send $3.95 to CSPI, 1501 16th St. NW, Washington, DC 20036.

HEALTH CURRICULUM. The American Cancer Society lends excellent health curricula designed for four elementary grade levels. Inquire locally about these series:

Take Joy. Describes the six systems of the body.

Health Network. Habits, attitudes, and decisions that form the network of good health.

SCRUBBY BEAR PROGRAM. The Department of Health and Human Services is sponsoring this program for preschools in a campaign to reduce infectious diseases through cleanliness education. Send for information about classroom materials: Scrubby Bear Foundation, 155 15th St. NW, Suite 500, Washington, DC 20005. Your local hospital may have a classroom program available. Your hospital may also distribute *Bobby the Healthcare Bear's Club News.*

EAT, THINK, AND BE HEALTHY. The Center for Science in the Public Interest nutrition curriculum guide for upper elementary grades. A good model for integrating nutrition study into the whole curriculum. The information conforms to today's federal guidelines for a healthy diet.

7

What Is Air?

Children become acquainted with hundreds of substances through their sensory encounters. They are intrigued by an invisible, ever-present substance that is not directly encountered by the senses until it affects something tangible. The activities in this chapter are designed to reveal these concepts:

- Air is almost everywhere
- Air is real; it takes up space
- Air presses on everything on all sides
- Moving air pushes things
- Air slows moving objects

In the experiences that follow, children will find air in empty containers and become aware of breathing air. They will feel air enclosed in something, see how it occupies space, and note how air pushes on everything. Also, they will enjoy making things for moving air to push.

Introduce the topic of air to children by holding up your hands, cupped together, suggesting that, "There is something important inside my hands. You may peek in, but you won't be able to see what it is. The important stuff is *invisible*! We'll be finding out more about it."

CONCEPT: Air is almost everywhere

1. What comes out of an empty can?

LEARNING OBJECTIVE: To develop a beginning awareness that air is present almost everywhere, even if unseen.

MATERIALS:

Empty metal can, #303

Nail, hammer

Dishpan

Water

Plastic aprons for participants

Splash catcher (Inflatable wading pool is fine. Use on floor, inflated to confine spills. Put pan of water in center of the pool. Children stay outside the pool.)

SMALL GROUP ACTIVITY:

1. Ask children to examine the can. Do they notice anything in it?
2. Punch a hole in the bottom with a hammer and nail. Let a child do it, if possible.
3. "If there is something in the can, could we push it out through this hole? Let's see."
4. Invert can; slowly push it into the water. Invite a child to hold his hand just above the nail hole. What does he feel being pushed out of the hole? A stream of air? "Could something real be in that empty can . . . something real that we can't see?"
5. Let the children experiment.

2. *Can we really empty an open container?*

LEARNING OBJECTIVE: To extend a beginning awareness that air is present almost everywhere, even if unseen.

MATERIALS:

Empty clear plastic shampoo bottles or tubes, uncapped

Downy feathers, milkweed fluff, or bits of tissue paper

SMALL GROUP ACTIVITY:

1. Ask children to examine the bottles or tubes. Do they notice anything inside?
2. "Point the bottle top toward your chin and squeeze it. Do you feel something? What? See if you can squeeze it out. Keep trying."
3. Offer a feather to each child. "If your bottle is really empty, nothing will come out to push the feather from your hand. See what happens if you point the bottle toward the feather." (This can be disorderly unless limits are set for a feather-blowing space.)

3. *Is there air inside us?*

LEARNING OBJECTIVE: To become aware of breathing air.

Group Activity: (Each child will need a paper tissue.) Suggest that all the children take a deep breath, shut their lips tightly, and pinch their nostrils shut as long as they can. "What happened? Why did you let go of your nose? What did your body seem to need so much that you had to let go? Let's try again."

"Our bodies need the part of air called oxygen. We need to breathe air into our bodies about every six seconds because it is used up so quickly inside of us." Guide the discussion to bring out the idea that all living things must have air to stay alive. Discuss the danger of climbing into boxes with heavy doors that cannot be opened from the inside. Give examples.

Give each child an unfolded tissue to hold near his or her nose and mouth. "Let's see what happens to the tissue when you bring air into your body and then when you push this air out. Try it with your lips closed; now with your lips opened a bit. Are there two ways for the air to come in and out? Let's see if we can tell where the air is going inside our bodies." Ask children to take turns lying on the floor or putting their heads on the table so others can watch their chests go up and down as they breathe.

Read *Whistle for Willie,* by Ezra Jack Keats. Comment about breathing air and box-play safety.

CONCEPT: Air is real; it takes up space

1. Can we feel the air inside a bag?

LEARNING OBJECTIVE: To develop a beginning awareness that air has substance.

MATERIALS:

Plastic sandwich bags (Not fold-top kind)*

Drinking straws

Masking tape

Covered wire closures

GETTING READY:

Prepare a blowing bag for each child participating by gathering the top of the bag to form a neck, inserting a straw in its neck, and securing the neck of the bag to the straw with tape.

SMALL GROUP ACTIVITY:

1. Ask "Is the table real? How can you tell? If your eyes were closed could you still tell that the table is real? Is feeling something a good way to know that it is real?"
2. Let children blow into the prepared bags. "What are you blowing into the bag? Look in the bag. Feel it. Do you think it is something real?"
3. Tie with wire closure so children can keep their bags full of air to feel.

*Balloons may be used instead of the sandwich bag and straws. However, the more transparent the container of air, the more apparent the concept that invisible air is taking up space inside the bag. We can feel air inside a container, we can see the container enlarge, but we cannot see the air. Children should leave the plastic bags at school so that younger children at home will not pick up the idea of playing with plastic bags.

2. Can a glassful of air push water away?

LEARNING OBJECTIVE: To extend the awareness that air has substance.

MATERIALS:

Clear plastic bowl (corn popper dome is fine) or 5" (12 cm) deep glass mixer bowl, if it can be safely used

Pitcher of water

Cork

Clear plastic tumbler

Splash catcher, aprons

GETTING READY:

Inflate wading pool splash catcher. Put bowl in the center.

SMALL GROUP ACTIVITY:

1. Put cork in the bowl. "The cork is touching the bowl now. What will happen if we pour water into the bowl? Let's see."
2. Fill half the bowl with water. "Notice if the cork is touching the bottom of the bowl now."
3. "Can someone put the cork back on the bottom?" Let children try to do this. "Will it stay if you let go?"
4. See if something in the tumbler will push water away so that the cork can stay on the bottom of the bowl. Invert tumbler and push straight down to the bottom. "What could be inside the glass pushing the water away? Do you see anything?"
5. Let the children experiment.

Brook can hold a cork under water without touching it.

Group Discussion: Ask the children to catch some air in their cupped hands. "Poke your nose into your hands. Can you smell the air? Peek in. Can you see it? Can you hear it? Try to taste it." (Children love to hang their tongues out for research purposes.) Ask those who have experimented how they know that air is real. Children are proud of their new information and are willing to repeat a discussion like this one. Next time, let a hand puppet be the skeptic who doesn't believe that invisible things can be real. The children will be happy to convince him otherwise.

3. *Can air in a bag push something over?*

LEARNING OBJECTIVE: To extend the awareness that air has substance.

MATERIALS:

Large, sturdy paper bag

Wire twist closure, or masking tape

Bicycle tire pump, or bellows-type air mattress pump

Heavy book or block

SMALL GROUP ACTIVITY:

1. Demonstrate the pump if children don't know what it is. Look at the air intake hole where air comes in to be pushed out by the pump. (Remove intake valve cap from bellows pump.) Feel air being pushed out.
2. "Do you think a bag full of air could push that book over? How could we find out?"
3. Insert the end of the pump hose into the bag. Tighten the closure, or simply hold bag firmly to hose while children take turns pumping air into it, pushing the book over. "Could air be doing that? Do you think air is real stuff, even though we can't see it?"

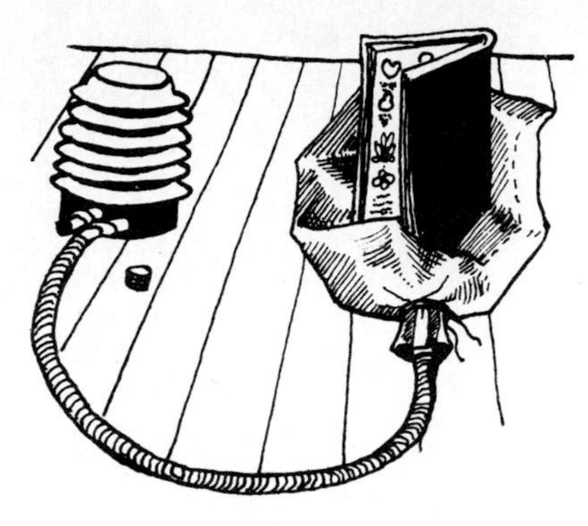

FIGURE 7–1

GETTING READY:

Gather bag top into a neck and fasten with a wire closure or tape.

Put bag on floor and stand book upright on it. Put the pump near the bag (Figure 7–1).

Note: Children love to demonstrate how strong they are. Using the pump is a good medium for this self-concept enhancement.

Read *The Ball That Wouldn't Bounce*, by Mel Cebulash (see Stories and Resources for Children).

CONCEPT: Air presses on everything on all sides

1. Can air keep water in a tube or straw?

LEARNING OBJECTIVE: To notice the effects of air pressure on liquids in a small tube.

MATERIALS:

Baster (kitchen tool)

Medicine droppers (plastic or thick glass tubes are safest)

Small containers of water for each child

Pint jar half full of water

Sponge, pan, and newspapers or splash catcher for cleanup

SMALL GROUP ACTIVITY:

1. Examine baster with children. Is the tube empty? Squeeze bulb to feel what comes out.
2. Put baster in jar of water. "What do you see when I push air out of the baster? Watch the tube when I slowly let go of the rubber bulb on top. What is happening?"
3. Hold up the filled baster. "What's happening to the water now? What could be keeping the water in? Perhaps something invisible is in the baster with the water."
4. Squeeze bulb to push out water. Release bulb to pull in air. "What's happening now? Try it."

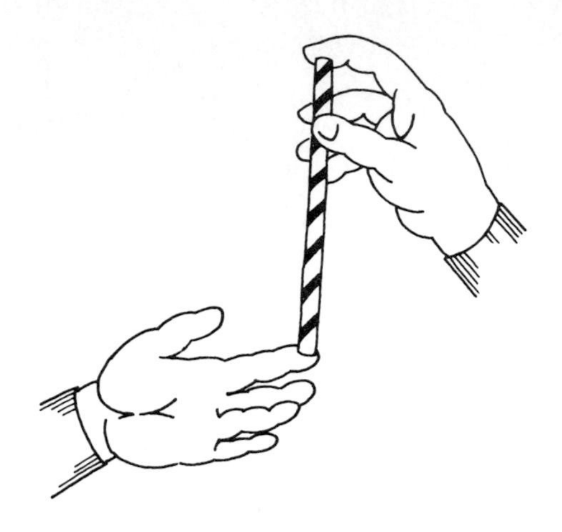

FIGURE 7–2

Drinking straws, cut in half

5. "Now let's try the same thing with straws." Insert a straw in water. Put your index fingers over the top and bottom of straw to lift it from the water. Keep finger on top and remove finger from bottom. Most of the water will stay in the straw. Why? (Air pressure keeps it in.) See Figure 7–2.
6. "Nothing can get into the straw from the top with my finger on it. Watch as I let some air push into the straw." Remove finger from top. Water is pushed out of the straw. Let children experiment.

Note: Some children may be satisfied that a dropper is filled with water if drops just cling to the outside. Help them hold the dropper tube in the water *while* releasing the bulb.

CONCEPT: Moving air pushes things

Introduction: Ask the children to wave their hands back and forth in front of their faces. "Can you feel something on your skin? When air is moving you can feel it. You don't notice it when it isn't moving." Explain that we feel air moving against us in the same way whether we are moving when the air is still or whether air is moving by itself (as wind) when we are still.

1. Does moving air push on us?

LEARNING OBJECTIVE: To experience the pushing effects of moving air.

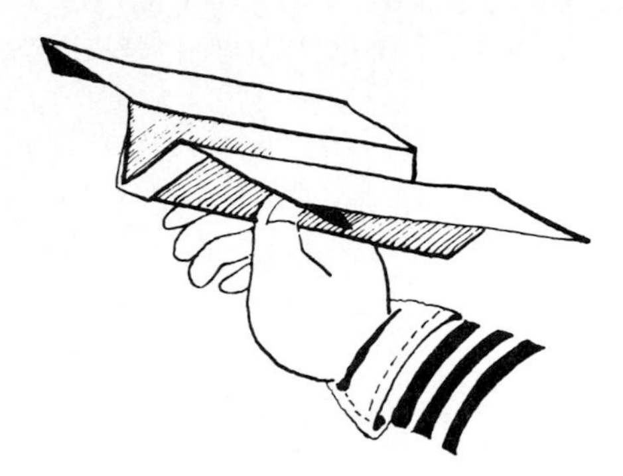

FIGURE 7–3

MATERIALS:

Sheets of newspaper

Used adding machine tape

Pinwheels or kites

OUTDOOR ACTIVITY:

Give children sheets of paper outside. "Will paper stick to you as you stand here? See what happens when you run as fast as you can. Do you have to hold the paper to press it to yourself? What could be holding it?" Let children enjoy running with paper tape streaming behind them. Try pinwheels standing still and running. Try the kite on a calm day and on a windy day.

Listen to "What Makes the Weather," sung by Tom Glazer in *Now We Know (Songs to Learn By)*. (Music references are listed in Appendix 1.)

2. *Can we make a glider drift on moving air?*

LEARNING OBJECTIVE: To notice that moving air can carry a glider launched by a child.

MATERIALS:

A sheet of typing paper for each child

GETTING READY:

Draw a lengthwise line down the center of the paper.

Draw parallel lines $1\frac{1}{2}''$ (4 cm) from the center line.

Draw a line $\frac{3}{4}''$ (2 cm) from the bottom across the paper.

SMALL GROUP ACTIVITY:

1. Show the children how to fold up $\frac{3}{4}''$ (2 cm) from the bottom of the paper. Continue to fold this amount 4 times.
2. With the cuff on the outside, crease the paper and the folded cuff lengthwise on the center line. Now fold back each side on the lines $1\frac{1}{2}''$ (4 cm) from the center fold, making wings.
3. Launch glider with cuff in front, holding the center, as shown in Figure 7–3. If the glider always nosedives, reduce the weight of the front end by unfolding one fold of the cuff. Also try to push the glider straight ahead.

Note: Primary-grade children may want to make more elaborate gliders. See *How to Have Fun Making Paper Airplanes*, by the editors of *Creative*.

The parachute works! It slowed the falling spool.

CONCEPT: Air slows moving objects

Group Discussion: Show the children a piece of typing paper and a small solid object like a rock or a wooden spool. "If I hold these two things up as high as I can reach and then let go of them, what will happen to them? (Wait for responses.) Do you think that both things will fall in just the same way? Let's find out." Repeat the action as needed to allow the children to carefully observe and report. "Which one fell more slowly? Can you think of anything beneath the wide paper that might have pushed against it and slowed its fall?" Mention that everything that moves above ground or above water must push air aside as it moves. Big things push more air aside than small things.

1. What does a parachute do?

LEARNING OBJECTIVE: To compare the ways air slows the fall of a weight with and without a parachute.

MATERIALS:

For each child:

1 wooden thread spool or stringing bead*

12″ (30 cm) square of closely woven, light fabric

4 feet (1.2 m) light string

Extra wooden spool

Large needle

SMALL GROUP ACTIVITY:

1. Allow children to help make parachutes if they can tie. Younger children can crayon designs on fabric before strings are tied.
2. If a safe stairway is nearby, let one child go up and reach through banisters to drop the extra spool and parachute. Compare results.
3. Fold parachute fabric and wind with the strings on spool. Let children throw them as high as they can into the air outdoors. Add weight to the spool if parachutes fail to open.

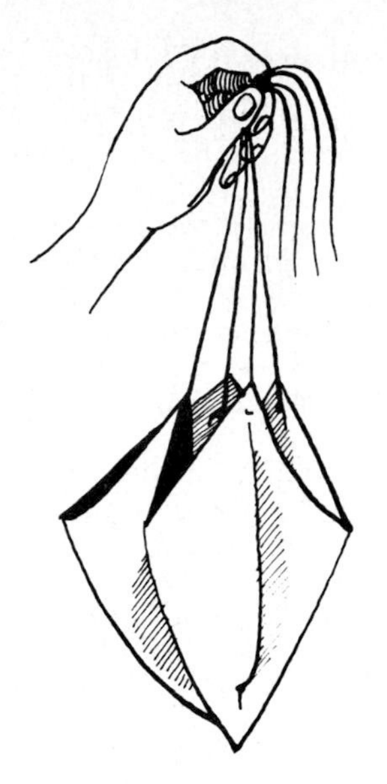

FIGURE 7–4

GETTING READY:

Sew a 12″ (30 cm) piece of string to each corner of fabric.

Hold parachute by strings, fabric corners even, as shown in Figure 7–4. Knot strings together 7 ″ (18 cm) from corners.

Tie spool to string ends.

*The spool must be heavier than the fabric. Tie on metal washers or nuts if more weight is needed.

PRIMARY INTEGRATING ACTIVITIES

Music (Resources in Appendix 1)

After children have finished clapping the rhythm of a song they enjoy singing, compare two styles of clapping. Listen to the sound made by clapping with palms open and flat, then to the sound made by clapping with slightly cupped hands. What is caught between the cupped hands to change the sound?

Sing this song to the refrain of "Goober Peas" (in *Folksong Festival*, Appendix 1):

Air, air, air, air,
air is everywhere.
We can't taste or see it,
but we know it's there.

Use appropriate tasting and peering gestures to add fun.

Stories and Resources for Children

BRANDT, KEITH. *Air.* Mahwah, NJ: Troll, 1985. Intended for elementary grade readers, portions of this book can be read aloud to pre-primary and primary-grade children. Paperback.*

BRANLEY, FRANKLYN. *Oxygen Keeps You Alive.* New York: Thomas Y. Crowell, 1971. Astronauts and scuba divers carry oxygen with them to survive away from air.

BRANLEY, FRANKLYN. *Air Is All Around You.* New York: Harper & Row, 1986. Simple air concepts and experiments, charming illustrations. Paperback.

BRIGHT, ROBERT. *Georgie and the Runaway Balloon.* Garden City, NJ: Doubleday, 1983. Georgie converts a handkerchief into a parachute to rescue a mouse with a runaway balloon.

CEBULASH, MEL. *The Ball That Wouldn't Bounce.* New York: Scholastic Book Services, 1972. A cast-off basketball needs some air before it will bounce.

CREATIVE, EDITORS OF. *How to Have Fun Making Paper Airplanes.* Pleasantville, NY: Reader's Digest Services, 1974.

ETS, MARIE HALL. *Gilberto and the Wind.* New York: Viking Press, 1963.

KEATS, EZRA JACK. *Whistle for Willie.* New York: Viking Press, 1964. Peter hides in a box that he can get out of easily. He also learns how to blow air out from his lips as a whistle.

LOBEL, ARNOLD. *Days with Frog and Toad.* New York: Harper & Row, 1979. Toad and Frog persist, in spite of discouragement, and finally are able to fly their kite in the wind.

MILNE, A. A. *Winnie-The-Pooh.* New York: E.P. Dutton, 1961, chapter 1. Read this after children have made parachutes. Ask whether a small balloon could really slow Pooh's fall.

NODSET, JOAN L. *Who Took the Farmer's Hat?* New York: Scholastic Book Services, 1970. Moving air is the culprit.

PODENDORF, ILLA. *The True Book of Science Experiments.* Chicago: Children's Press, 1972.

REY, H. A. *Curious George Gets a Medal.* New York: Scholastic Book Services, 1957. George drifts to earth under a parachute after a space flight.

SABIN, FRANCENE. *Whales and Dolphins.* Mahwah, NJ: Troll, 1985. All mammals breathe air, even those that stay underwater for long periods without surfacing for air.

SCARRY, RICHARD. *Great Big Air Book.* New York: Random House, 1971. The air we breathe, flight principles, parachutes, moving air and its uses, pollution, and the absence of air on the moon are all part of this learn-while-laughing book.

SELSAM, MILLICENT. *Plenty of Fish.* New York: Harper & Brothers, 1960. Willie discovers that fish and people get air in different ways. An amusing story.

WYLER, ROSE. *Science Fun With Toy Boats and Planes.* New York: Simon & Schuster, 1986. Includes good information about the lifting power of air streams, and paper glider construction.

Poems (Resources in Appendix 1)

The wind is a favorite topic in poems for children. The following are recommended:

- In *Poems Children Will Sit Still For,* by de Regniers:
  "Who Has Seen the Wind," by Christina Rossetti.
  "Wind Song," by Lillian Moore.

*Starred references, written at the young child's level of understanding, can help teachers with minimal backgrounds in science expand their knowledge base.

- In *Now We Are Six,* by A.A. Milne:
  "Wind On the Hill."
- In *Poems to Grow On,* by Jean Thompson:
  "The Kite," by Harry Behn.

Fingerplays

Seeds with silky wings are scattered by moving air (see chapter 4, Plant Life). Recall the milkweed blowing that children may have enjoyed in plant life experiences, or try to find a milkweed pod to open outdoors now if it is a new idea.

Baby Seeds

In a milkweed cradle (cup hands)
Snug and warm
Baby seeds are hiding (make yourself small)
Safe from harm.
Open wide the cradle (open hands)
Hold it high.
Come Mr. Wind (blow in the hands)
Help them to fly.

—AUTHOR UNKNOWN

Art Activities

1. ***Easel Painting.*** Cut newsprint into the shapes of things that move on the air such as birds, balloons (tape a dangling string to the back), butterflies, and kites. Let each child choose a shape and paint with a design of his or her own creation.
2. ***Collage.*** Use airborne nature materials such as feathers (to represent birds); milkweed fluff and seeds; and maple tree seeds. Cut some actual air bubbles from sheets of plastic packing material to glue on the collage.
3. Make paper fans to move air. They may be accordion-pleated sheets of paper which the children decorate or they may be small paper plates stapled to Popsicle sticks.
4. Make simple gliders to decorate with crayon designs.
5. Use air pressure with medicine droppers as an art medium. Let children make designs by dropping diluted food coloring onto absorbent paper with the droppers. A fleeting, fascinating art form—snow painting—also makes use of droppers and food coloring. Gather a big bucket of snow. Store it on the windowsill. Quickly pack individual plastic meat trays with snow, ready to receive the drops of color. Have another bucket ready for the sloshy end results.
6. Do blow painting with plastic straws. Children can use straws and air pressure to pick up tempera paint from jars and drop it on smooth-surfaced paper. The puddles of paint are then changed in shape by blowing on them through the straws.

Dramatic Play

Service Station. Contrive a gasoline pump from an oblong carton and a length of garden hose. Help the children create a car with chairs, hollow blocks, and a paper-plate steering wheel (or a real one). Add an old inner tube and a bicycle tire pump for the repairman to inflate. A ring of keys and pliers add fun.

Boat. Help the children create a block and plank boat, with two planks angled together to form a prow. Let them blow up a plastic ring or pump up the inner tube for a life preserver.

Plane. Angled planks form the nose of the plane, hollow blocks form the tail and sides, two more planks stretch out to form the wings. Children who have flown can contribute ideas about oxygen and air to the play.

Housekeeping Play. Try to borrow a child-size inflatable chair to add interest to this play. Let the children beat up a detergent suds "soufflé" at the sink with a wire whisk or eggbeaters. Comment on the amount of air their strong muscles are beating into the dish.

Other Air-Related Play

1. Blow soap bubbles, using straws in cups of water and detergent. For younger children use baby shampoo, which won't irritate eyes, instead of detergent.
2. Let children take turns pumping up a sturdy air mattress to use as a tumbling mat for somersaults.
3. Pump up an inner tube to use across the sets of two facing chairs as a target for foam ball tossing.
4. Let moving air make paper dance. Cut a spiral from a sheet of paper to make one long, curving strip. Dangle it from the ceiling by attaching it to a long thread. Let children fan the air with a card to make the spiral twirl. Create a fantasy effect by dangling several spirals. Circulating air will keep them gently moving.

Food Experiences

If an electric mixer is available, make whipped milk topping. State that the beater is stirring many tiny air bubbles into the mixture. Watch it change and expand.

Combine: $\frac{1}{2}$ cup instant dry skim milk
$\frac{1}{2}$ cup *cold* water

Beat at highest speed for 4 minutes.

Add: 2 tbsp. sugar
$\frac{1}{2}$ tsp. vanilla

Beat at highest speed until mixture stands in peaks, then about 4 minutes longer. Serve a mound of whipped milk to each child. Edible science is impressive. Children really remember that air was part of the tidbit they enjoyed eating.

Thinking Games: Logical Thinking and Imagining

Classifying Game. Find pictures of things that move with the help of air (sailboats, planes, kites, balloons, birds, bats, butterflies, and so forth) and pictures of things that move without the help of air (trains, steamships, motor boats, caterpillar tractors, roller skates, and so on). Mount the pictures on cardboard and put them in a box with *yes* or *no* marked inside the lid and the bottom section, respectively. Use the pictures as a game with the whole class or as a quiet table game for a few children. (Pictures of cars and other things using inflated tires could be classified either way. Air does not push a car along, but a car with a flat tire certainly won't go very far. Let the children decide the issue after discussing it.)

After singing the "Air" song, let the children enjoy thinking of funny places where air creeps in. Start off the game with places like: inside your shoe, your ear, the holes in Swiss cheese, beneath a mosquito's wings . . .

Field Trips

Airport Visit. If this is possible, keep the science focus simple. There will be too many exciting things to see to make lengthy explanations of flight principles bearable for the children. Be sure that any adult who offers a guided tour of the facility understands this.

Service Station Visit. Make arrangements with the station manager beforehand to have a demonstration of the tire changer and to allow children to feel air from the tire pump. This can be done in connection with a study of simple machines (Chapter 13) if the station is too far away to warrant two trips.

Creative Movement

1. Let children pretend to be balloons that are blown into large shapes. Suggest that they make themselves into flat, limp shapes on the floor. "Now I'm blowing, and blowing, and blowing some more. And you are getting bigger and bigger." When the shapes have been stretched as large as they can be, suggest that you will pretend to prick each imaginary balloon shape. Each shape collapses into a limp piece of rubber on the floor as the air goes out of it.
2. Children can move like birds or butterflies soaring on moving air. They might enjoy holding a feather in each hand to become a bird with outstretched-arm "wings."
3. The library may have a record-lending service including an RCA Victor Dance-A-Story record by Anne Lief Barlin entitled *Balloons* (see Appendix 1).

It will offer an example to those who are new to creative movement, indicating how spoken ideas guide the movement.

4. Some excellent ideas for related creative movement can be found in *Creative Movement for the Developing Child*, by Clare Cherry (see Appendix 1).

SECONDARY INTEGRATION

Keeping Concepts Alive

1. Make casual comments or inquiries about the function or presence of air when suitable opportunities arise. When outdoors with a class, observe planes or birds flying overhead, or leaves, seeds, or litter blowing past.
2. Suggest that children fill their mouths with a giant air bubble when the class needs to walk quietly through school corridors. Puff up your own cheeks to model this funny way to stay quiet. It's more positive than saying, "Be quiet!" Air takes up talking space.
3. Use medicine droppers to water tiny seedlings, give moisture to insects in jars, and for art projects. If necessary, remind children to push the air out of the dropper while it is still in the liquid. Can they see the air bubbling into the liquid?
4. If you need to open a vacuum-sealed container, such as using a punch-type opener with a can of juice, let the children listen closely for the hiss of air rushing into the can. When using the bellows-type step pump, show the rubber cap on the air intake valve. Push the air out so the bellows are flat. Remove the cap so that air rushes in with a hiss and the bellows expand dramatically.

Relating New Concepts to Existing Concepts

1. Relate water concepts to air when investigating evaporation. Water droplets, or *vapor*, become part of the air. Speak of fog and clouds as very wet air.
2. Relate plant propagation (seed scattering) to moving air.
3. Relate sound to air concepts. Sound travels through air; moving air causes some things to vibrate and make sounds; air vibrates when enclosed in a column like a flute or a whistle (see chapter 14).
4. Point out the crucial role of air in ecological relationships. Air is one of the nonliving substances that all living things depend on to stay alive. Air that is spoiled by pollution makes problems for all living things. There is no way to make new air. It is used over and over again, so people must find ways to keep air clean.

Family Involvement

A newsletter to parents can suggest ways to support and add to the air experiences at school. When families are out together, they can watch for soaring birds, planes, and clouds being pushed along by moving air. They can point out how air is used in their homes to keep people comfortable: heated in winter, passed through coolers in summer. Notify families if there are exhibits in your area where children can see model or real gliders, small planes, windmills, or weather vanes.

EXPLORING AT HOME*

Impressive Pressure! Enjoy sharing this surprising air pressure experience with your child. You'll need quart-size plastic bags (Reuse any handy bag that doesn't have any holes—not even a pinhole), a widemouthed pint jar, and a rubber band. Try this by yourself first, then let your child do it. Put your hand inside the plastic bag, then insert the bag into the jar. Smooth it against the *inside* of the jar. Fold the rest of the bag back over the *outside* of the jar. Fasten the bag with the rubber band at the neck of the jar. Now, reach in and try to pull the bag out of the jar. (It won't pull free. Why? Because more air is pushing down inside the bag than between the outside of the bag and the jar.) Cut a tiny hole in the bag and try it again. (The bag pulls free as more air gets into the space between the outside of the bag and the jar. Then the air pressure is just as strong on the inside as on the outside of the bag.) Let your child experiment with suction cups you might have in your home, such as kitchen cup hook holders or a "plumber's friend."

RESOURCES

AMERY, HEATHER. *The KnowHow Book of Experiments.* London: Osborne Publishing, 1979. Another air pressure experiment is included in this collection. Paperback.

BROEKEL, RAY. *Experiments with Air.* Chicago: Children's Press, 1988. Discover the existence of air in soil, the nature of a jet of air, and play an easy wind instrument. (Photographs show children handling materials that should be used only under adult supervision, unfortunately.)

LEVENSON, ELAINE. *Teaching Children About Science.* Englewood Cliffs, NJ: Prentice-Hall, 1985.

VAN CLEAVE, JANICE. *Chemistry for Every Kid: 101 Easy Experiments That Really Work.* New York: John Wiley, 1989.

WHITE, LAURENCE. *Science Toys and Tricks.* New York: Harper & Row, 1990. The impressive upside-down glass of water experiment demonstrating air pressure, and a simple aerodynamics principle are in this collection. Paperback.

*This suggestion may be duplicated and sent home to families.

8

What is Water?

All living things require water to survive. Water is also splendid for pouring, splashing, and spraying. Most children enjoy learning about this vital part of our environment. The experiences in this chapter will explore the following concepts:

- Water has weight
- Water's weight helps things float
- Water goes into some things, but not others
- Some things disappear in water; some do not
- Water goes into the air
- Water can change forms reversibly
- The surface of still water pulls together

It is hard to imagine sparkling, transparent water having weight. Once children recognize that water does have substance, they understand more readily how water can hold things up.

In the following experiments children experience the weight of water, note that water is heavier than air, and explore buoyancy. Other experiences deal with water absorption and repellency, objects that dissolve in water, evaporation, condensation, freezing, melting, and surface tension of water.

CONCEPT: Water has weight

1. Does water feel heavy?

LEARNING OBJECTIVE: To experience firsthand that water has weight.

MATERIALS:

Outdoors when it's warm:

Small wading pool

Large bucket

Large pitcher or milk cartons

Water

SMALL GROUP ACTIVITY:

1. Pass pitcher of water for children to dabble fingers in. "Do you think this clear-looking water could be heavy? Let's find out."
2. Let children take turns standing in the pool holding the empty bucket.

For indoor use add:

Rubber boots

Plastic smock

GETTING READY:

If outdoors, have children remove shoes and socks and roll up trouser legs.

If indoors, have children wear boots and smock.

3. Let other children pour water into the bucket until it becomes too heavy to hold. Conclusions about the weight of water are easily formed and communicated by this direct investigation.

2. *Do things weigh the same when dry and when damp?*

LEARNING OBJECTIVE: To observe that the weight of a material increases when water is added to it.

MATERIALS:

Part 1

Inexpensive postage or calorie-counting scale

Matching set of small cups such as film canisters, spray can tops, etc.

Small pitchers or half-pint milk cartons

Water

Sponges, newspapers

Part 2

Tiny scoops

Sand

Pan for wet sand

GETTING READY:

Spread newspaper on a table.

Cover postal rate finder, etc., with masking tape, so the indicator points only to ounce numerals.

Make a simple chart for children to keep weight records of dry and wet material.

SMALL GROUP ACTIVITY:

1. Put an empty cup on the scale. "What does a scale do? Let's see what happens if we pour water into the cup. What do you think it means when the pointer moves down? Which is heavier, a cup of air or a cup of water?
2. Put dry sand in two matching cups. Weigh each to be sure they are equal.
3. "Try pouring a bit of water in the sand cup on the scale. Where is the pointer now? Why do you think it moved from 2 to 3?" Compare with weight of the cup of dry sand. "Do they weigh the same now?"
4. Let children measure, weigh, and add water. Some may wish to try each set of cups to verify results.
5. Let children record dry and wet sand weight on the chart.
6. Read the chart results to the whole class after the experiment. What was added to the damp sand that made it heavier?

CONCEPT: Water's weight helps things float

1. Will some things float while others sink?

LEARNING OBJECTIVE: To compare which objects among a collection varying in dimension and weight will be supported by water and which will not.

MATERIALS:

Dishpans of water

Small wading pool

Plastic aprons

Assorted objects: rocks, sticks, bath toys, corks, washers. Try to find like objects of different materials: ping-pong ball/golf ball; toy plastic key/metal key

Old kitchen scale, postage scale, or balance

Two trays

GETTING READY:

Place pans of water in the wading pool.

Tape card marked *float* on one tray and *sink* on the other.

SMALL GROUP ACTIVITY:

1. Let children enjoy relaxed play with the materials.
2. "Does the water hold some things up on top? All things? Why do you think this ball stays on top and that ball sinks down?"
3. Encourage children to weigh objects on the scale to confirm that similar objects can differ in weight and thus can act differently in water.
4. Put out *sink* and *float* tray. Suggest classifying objects and placing them in correct trays.

2. Can air help some things float?

LEARNING OBJECTIVE: To discover whether air inside an object helps the object float on water.

MATERIALS:

Dishpans of water

Small wading pool

Plastic aprons

Foot-long (30 cm) pieces of clear garden hose

2 corks to fit each hose

Many small plastic bottles (prescription vials *with* caps)

SMALL GROUP ACTIVITY:

1. Remove one cork from a hose. Submerge it in water. Let it fill almost to the top with water. Replace cork. "What is inside the hose now? Is air or water on top? Which do you think will be on top if the hose is turned upside down? On its side? Try to find out."
2. Drop two capped bottles on the water. "What might happen if you take the cap off one bottle? Do you think it will still float? Find out. What could be in the bottle that still has a cap on it?"

Phillip feels and sees that cloth sprayed with water won't keep his arm dry.

GETTING READY:

Fit corks securely into hose ends.

Place pans of water in the dry wading pool.

Read *The Story About Ping,* by Marjorie Flack.

CONCEPT: Water goes into some things, but not others

1. What things will absorb water?

LEARNING OBJECTIVE: To observe which materials among a collection will absorb water and which will repel water.

MATERIALS:

Plastic aprons

Dropper-top medicine bottle for each child

Tiny funnel and baster

Plastic meat trays

SMALL GROUP ACTIVITY:

Part 1

1. "What happens when you put drops of water on different things in your tray?" Listen to the children's ideas. Define the term *absorb.*
2. Give a small cube of dry sponge to each child. Put in a dry place. "Watch closely; something will change fast." Squeeze water from baster onto the sponge.

Test materials: sponges, wood, cotton, paper towels, tissues, dried clay, stones

6'' (15 cm) squares of old cotton fabric

Plastic sheeting or rubberized fabric pieces

Waxed paper, scraps of leather

Smooth feathers

Spray bottle

Water

GETTING READY:

Let children fill bottles with water.

Arrange a tray of test materials for each child.

Part 2

1. Move from child to child with a piece of plastic, dry fabric, and the spray bottle.
2. Drape cloth over child's arm. Spray with water. "What do you feel?" Repeat with plastic. "Which material would you want for a raincoat? Why?" (This firsthand experience with water repellency intrigues children.)
3. Give each child a feather to hold while you carefully place *one* drop of water on the feather. Listen to the responses. Talk about birds' staying dry in the rain. Let children experiment with water drops, feathers, and leather scraps.

Read *Wet and Dry,* by Seymour Simon.

Group Experience: 1. Put some dried beans into a plastic screw-top jar. Mark the level of beans with a wax pencil. (Use less than half a jar of beans.) Let each child squeeze a dry bean and put it in the jar. How did it feel? Pour water to cover the beans. Cap the jar. Set aside a few dry beans for comparison. The next day, check the amount of space the beans take up in the jar. What happened? Where is the water? Compare soaked and dry beans.

2. Weigh a dry sponge on a kitchen scale. Drop it in a bowl of water. Ask children for predictions about the weight of the wet sponge. Reweigh the sponge. Let the children decide what made the weight difference. Squeeze the water out over a dry bowl. Weigh it again.

3. Use the term *absorb* whenever a spill must be wiped up from now on. Let the children use the sponge to absorb spills. They feel more in control of the situation if they can undo what they have done accidentally.

CONCEPT: Some things disappear in water; some do not

1. Which things dissolve in water?

LEARNING OBJECTIVE: To observe which materials among a collection will dissolve in water and which will not.

MATERIALS:

Muffin pans and plastic ice cube trays with separated molds, or plastic prescription vials

SMALL GROUP ACTIVITY:

1. "See what happens when you put a little salt in one of your pans of water. Stir it. Can you see it? Feel it? Where is it?"

Pitcher of water

Assorted dry materials: salt, sand, cornstarch, flour, fine gravel, seeds, cornmeal, etc.

Spoons for dry materials

Salad oil

Small screw-top bottle

Plastic aprons

Newspaper

Sponge

Sticks for stirring

Cleanup bucket

GETTING READY:

Spread newspapers on work table.

Half fill pans with water.

2. "Dip a finger in the pan. How does the water taste? The salt is still there, but it is in such tiny bits now it can't be seen. It *dissolved* in the water."
3. "Try the other materials; put each in its own pan of water. Find out which ones dissolve."
4. Half fill the bottle with water. Add some oil. Cap securely. Let the children shake it. Does it seem to dissolve? Let it stand a while. Where is the oil now?

Group Experience: Mix some sand, dirt, and gravel with water in a pint jar. Let it stand undisturbed for a day or more. Check the jar. Is the water still muddy looking? Did the sand and dirt really dissolve? Which is on the bottom?

CONCEPT: Water goes into the air

1. Will warm air and moving air take up water?

LEARNING OBJECTIVE: To observe the results of air taking up water and to participate in hastening the process.

MATERIALS:

Two trays

Pans of water

Paper towels

Blackboard or other dark, smooth surface

Cardboard or paper fans

Hand-held hair dryer

Paper

SMALL GROUP ACTIVITY:

1. "If we wet some towels, squeeze them, and spread them out on these trays, what do you think will happen to them? Let's see." Put one tray in a sunny or warm place, the other in a cool, darker place. Check them in half an hour.
2. Meanwhile show children the hair dryer; feel warm air coming from it. "What do you feel? This warm air can help us understand what is happening to the towels."

GETTING READY:

Check safety of the dryer cord position.

Let children make folded paper fans.

If possible, use the hair dryer close to floor level to minimize tripping on the cord or dropping the dryer.

3. Let children dip a finger in water, then trace their names on the blackboard. "Blow warm air on your wet name. Watch closely to see what happens to the water. Where did it go? Only air touched it. The water has gone into the air in such tiny drops that it can't be seen anymore. When air takes up water, we say the water *evaporates*.
4. "Write your name with water on the blackboard again. Fold a paper fan and wave it near your name to make a breeze. Where does the water go? Let's see what happened to the paper towels. What happened to the wetness?"

Read *The Rain Puddle*, by Adelaide Holl.

Group Experience: Early in the morning put measured amounts of water in three pie pans. Place pans where children can see but not touch them. Have one in the sunlight, one in the shade, and one close to a heat source. Check several times during the day to note changes. If the weather is humid, you may need to continue to check the next day. What happened?

CONCEPT: Water can change forms reversibly

1. Will temperature change make water change?

LEARNING OBJECTIVE: To observe how changes in temperature transform water from liquid to solid and from solid to liquid.

MATERIALS:

Water

Two identical shallow plastic bowls. (Not ice cube trays. In some parts of the country children may believe that ice is formed only in cube shapes.)

Access to a freezing compartment or, better yet, to freezing weather

SMALL GROUP ACTIVITY:

1. Early in the morning let children fill bowls almost to the top with water. Help them deliver one bowl to the freezer or to the near-zero outdoor location. Leave the other bowl in the room (out of spilling range) for an hour or more.
2. "Do you think both bowls of water will stay the same or will one change? Let's check."
3. Bring ice back to the room when it is solid. Dump it out onto a pie pan for all to see.
4. "What do you think will happen if we leave the bowl of water in the cold place? Let's find out." Repeat the activity if the children request it.*
5. Bring snowballs or icicles indoors to melt in a pan, when possible.

Read *The Snowy Day*, by Ezra Jack Keats.

*Some children may have had enough experience with water and ice to predict reversibility of this change in water form. The transformation from water to vapor through temperature change is less familiar. Both forms of reversibility may need repeating several times to begin to clarify cause-and-effect ideas for less experienced children.

FIGURE 8–1

2. *Can water change to vapor; vapor change to water?*

LEARNING OBJECTIVE: To observe how changes in temperature transform water from liquid to vapor and from vapor to liquid.

MATERIALS:

Vaporizer, baby bottle warmer, or facial sauna heating unit

Rectangular cake pan

Blocks, bricks, or books

Ice cubes

Water

GETTING READY

Arrange two stacks of blocks on either side of the heating unit, high enough to support the pan above the boiling water.

Check water heating unit cord and locate it where it can't possibly be tripped over.

Do not let children do this without close adult supervision of the heating unit.

SMALL GROUP ACTIVITY:

1. Show children the heating unit (Figure 8–1). "I will plug this in, then add water that will get very *hot*. It will change to invisible droplets called *vapor*. It will look like a pretty cloud, but it will be *burning hot! It must not be touched!*"
2. Connect the unit and add water in the amount stated on the unit label. "This hot vapor is steam. Clouds and fog are vapor, too, but they aren't as hot. Clouds, fog, and steam are all wet air."
3. Let children fill the pan with ice. Place the pan over boiling water, resting it on block stacks. "Let's see what happens when wet air gets very cold."
4. Watch droplets collect on the bottom of the pan and fall. "Water changed to vapor when it got hot. Vapor changed to water when it got cold. The change is called *condensation* when vapor becomes water."
5. Comment that the droplets of water in clouds, fog, or steam are so tiny that they can float in the air. When droplets come together in cold air, they become too heavy to float. They fall down as raindrops.
6. Go slowly with this. Listen to children's ideas. Don't expect them to verbalize everything that happened. Do again, if possible.

Read *Georgie and the Noisy Ghost*, by Robert Bright.

Group Activity: Ask the children to breathe slowly into their cupped hands. Does their breath feel warm or cold when it comes from their bodies? Pass out small mirrors or foil pans for children to exhale on. Are these surfaces cool? Ask for predictions about the outcome of breathing on the cool mirror or metal. Find out what happens. Can the children see results? "How does the cloudy place feel? Is it wet or dry? Try it again. Feel the cloudy place, and then feel a place that wasn't breathed on. Why do we feel wetness?"

CONCEPT: The surface of still water pulls together

1. Are drops of water curved?

LEARNING OBJECTIVE: To notice the curved form taken by drops of water and the adhesion property of water drops.

MATERIALS:

Water

Small funnel

Small bottles

Medicine droppers

Baster

Waxed paper or formica-topped table

Plastic aprons

Spoons, Popsicle sticks

Sponge to *absorb* spills

GETTING READY:

Fill individual bottles with water. Children like to do this with a funnel and small pitcher. Let them help, if possible.

SMALL GROUP ACTIVITY:

1. Give each child a square of waxed paper. "What will happen if a bit of water is squeezed from the dropper onto the paper? Find out. Did it make a flat splash or a rounded drop? Watch closely."
2. Suggest seeing what happens when many drops are made close to each other and when drops are made with the baster.
3. Talk about the outside edges of the drops pulling together to make an invisible "skin." This "skin" is not very strong.
4. Suggest gently dipping the spoon handle or stick into a big drop watching closely. Notice how the water clings as the spoon is slowly pulled away.

2. Can surface tension hold some weight?

LEARNING OBJECTIVE: To explore the nature of surface tension.

MATERIALS:

Clear glass or plastic jar, about 4" (10 cm) in diameter

Water

Syringe-type baster

Plastic aprons

Sponge and newspapers to *absorb* spills

Bucket for cleanup

Talcum powder, paper, uncooked spaghetti, aluminum foil, waxed paper

SMALL GROUP ACTIVITY:

1. Fill one or two jars almost to the top with water.
2. While children watch with heads close to the table, slowly add drops of water from the baster until the jar is full. Ask, "What might happen if more drops are added?" Find out, drop by drop. Did the surface pull more drops into a curve of invisible "skin?"
3. Hold a piece of spaghetti with thumb and forefinger. Gently rest it on the surface. Now push it through the "skin" vertically. What happens? (Only dry pieces rest on the "skin.")

3. Does soap break water's surface tension?

LEARNING OBJECTIVE: To cause surface tension to dissipate, using soap.

MATERIALS:

Pepper or talcum powder

Pitcher of water

Small jar

Spoon

Slivers of soap

Shallow foil pans

Liquid detergent

Plastic aprons

Newspaper, bucket, and sponge for cleanup

GETTING READY:

Pour 1'' (2.5 cm) of water in foil pans.

Half fill with water.

SMALL GROUP ACTIVITY:

1. "If we stir pepper (or talcum) and water together in this jar, will they mix? Find out."
2. "Will they mix if we gently sprinkle pepper on top of the water? Try it with the pans of water."
3. "What could be keeping it on top of the water?" (surface tension, the invisible "skin")
4. "Let's see what happens when we touch the surface with a bit of soap."
5. Let the children experiment, changing water as needed. Try adding a few drops of detergent instead of soap. What happens?

PRIMARY INTEGRATING ACTIVITIES

Music (Resources in Appendix 1)

1. It's fun to be vigorously involved in singing "Row, Row, Row Your Boat" by joining hands and touching feet with a partner to pull back and forth. It is good to explain that water holds boats up, also.
2. Sing "The Eency Weency Spider" after the evaporation experiences with an adaptation in lyrics. Change "and dried up all the rain" to "and evaporated the rain." Look for rain spouts near the school which children may watch after a rain.
3. Sing the old nursery song "One Misty, Moisty Morning."

Misty Moisty was the morn, Rainy was the weather.
When I met an old man, Dressed all in leather.

After experimenting with absorption and repellency, discuss with the children whether the old man was dry inside his leather clothes.

4. Sing this song after experimenting with forms of water (to the tune of "Twinkle, Twinkle, Little Star"):

Water, water from the stream
When it boils it turns to steam.
Water, water is so nice;

Freeze it cold, it turns to ice.
Cool the stream; warm the ice
It's water again, clear and nice.

Math Experiences

1. Encourage the children to count aloud the number of stirs each child takes as drink mixes or Jello are dissolved. Let the children be the clock watchers to remind you when the Jello should be solidified. If you have a play clock, set the hands to the appointed time. The children will have a means of checking the hand position of the real clock.
2. Let the children record the weight of materials being compared in dry and wet states. A discussion of the results could bring in more than/less than comparisons. "Are 4 ounces more than 2 ounces?"
3. Provide a set of measuring cups or three sizes of paper milk cartons for casual use by children playing with water. Ask how many small containers of water it takes to fill a cup or carton this size, and so on.

Stories and Resources for Children

ARVETIS, CHRIS, & PALMER, CAROLE. *Why Does It Float?* Chicago: Rand McNally, 1983. Buoyancy principles are explained through experimentation.

BAINS, RAE. *Water*. Mahwah, NJ: Troll, 1985. Portions of this reference, such as the water cycle, can be read aloud to pre-primary children.*

BREWER, MARY. *What Floats?* Chicago: Children's Press, 1976. Written in the voice of the young child making the discoveries, this is one of the clearest presentations of the water displacement concept for young children.*

FLACK, MARJORIE. *The Story About Ping*. Seafarer ed., New York: Viking Press, 1970. In this classic story, a barrel tied to his back keeps a houseboat boy afloat after he tumbles into the Yangtze River.

GOLDIN, AUGUSTA. *Ducks Don't Get Wet*. New York: Harper & Row, 1989. Try the duck's waterproofing system described here.

HOLL, ADELAIDE. *The Rain Puddle*. New York: Lothrop, Lee & Shepard, 1965. Evaporation of a puddle confuses some already befuddled animals.

KEATS, EZRA JACK. *The Snowy Day*. New York: Viking Press, 1962. What happens to the snowball in Peter's pocket?

LOBEL, ARNOLD. *Owl at Home*. New York: Harper & Row, 1975. Owl lets freezing air into his house. His soup freezes in the bowl. Later, when owl starts a new fire in the fireplace, his soup thaws.

MILGROM, HARRY. *ABC Science Experiments*. New York: Macmillan, 1970. Several water experiments are included in this picture book for very young children.

MILNE, A. A. *Winnie-The-Pooh*. New York: E. P. Dutton, 1961, chapter 9. "In Which Pooh is Entirely Surrounded by Water." Pooh and Christopher Robin use unorthodox means of floating through flood waters.

REY, H. A. *Curious George Rides a Bike*. New York: Scholastic Book Services, 1973. George forgets to deliver newspapers because he is so absorbed in making newspaper boats and floating his fleet downstream.

*Starred references, written at the young child's level of understanding, can help teachers with minimal backgrounds in science expand their knowledge base.

SIMON, SEYMOUR. *Wet and Dry*. New York: McGraw-Hill, 1969. This is an appealing book of water experiments designed for very young children.

SIMON, SEYMOUR. *Soap Bubble Magic*. New York: Lothrop, 1985. After enjoying bubble play, children are skillfully guided to discover the effects of surface tension, air concepts, and optical effects. Children are encouraged at the end of the book to continue to think about what they see and do.*

ZUBROWSKI, BERNIE. *Bubbles*. Boston: Little, Brown, 1979. Directions given for making wonderful bubbles: bubbles inside of bubbles, giant bubbles, geometric bubbles, and more.

Poems (Resources in Appendix 1)

From *Poems Children Will Sit Still For* read "Dragon Smoke," by Lilian Moore, about the moist air exhaled from our lungs becoming visible vapor in cold weather.

Fingerplays

Here is a fingerplay about evaporation:

In soapy water	(scrubbing motion)
I wash my clothes,	
I hang them out to dry.	(pantomime)
The sun it shines	(hands form circle)
The wind it blows,	(wave arms, sway)
The wetness goes into the sky.	

Art Activities

Ice Cube Painting. Use finger paint and glossy paper, but do not wet the paper as you would to prepare for finger painting. Instead offer the children ice cubes with which to spread and dilute the paint while making designs. Children who have watched ice melt like to apply the effect in something they create. (Suggested by Lilian Dean.)

Tempera Painting. Let children watch or help you mix dry tempera paint with water. When the finished paintings are hung to dry, use the term *evaporate* to explain how this drying occurs.

Dramatic Play

Indoor Camping. Devise a partial tent by securing a blanket over a climbing box, or arrange it lean-to fashion from a high shelf to the floor. Make a fishing pond by placing a dishpan of water in an empty wading pool. Provide a stick and string fish pole. Tie a cork using 6'' (15 cm) from the end as a bobber above water and a metal washer on the end of the string as a sinker.

Housekeeping Play. Wash doll clothes. If possible hang some to dry in the sunshine, some in a shady location, and some indoors. Compare drying times.

Indoor Water Play. Place two or three dishpans of water inside an empty wading pool. Provide plastic or metal pitchers, plastic funnels, and lengths of tubing for play. Mark several plastic baby bottles with tape and numerals to indicate levels to which the bottles can be filled with water.

Outdoor Warm Weather Water Play. Barefooted children can transfer water from buckets to the wading pool, using a plastic hand pump from the automotive supply section of a variety store. Try putting a bucket on a pile of hollow blocks next to the wading pool for experimentation with a garden hose siphon. Read *Curious George Rides a Bicycle,* by H. A. Rey. Then help children make newspaper boats to float in the wading pool, following the techniques George uses in the story.

Bubble Making. Choose some of these ways to have fun with surface tension:

- Use commercial bubble solutions and ring-tipped wands to wave or blow bubbles. Replenish the solution with a concentrated mixture of liquid detergent and water. If available, try the commercial set of giant rings and multiple bubble template. Use these outdoors.
- Provide straws and cans of detergent/water solution for children to blow *into.* This is easier for children who aren't able to cope with blowing bubbles into the air.
- Try fat drinking straws as pipes for blowing detergent/water solution bubbles into the air. Show the children how to dip the end of the straw into the solution to let a film collect across the bottom of the straw. Suggest holding the straw slightly downward to avoid dripping solution into the mouth.
- Try using soft plastic funnels to blow giant bubbles. (Have several ready and be prepared to wash mouthpieces before sharing.) Put the funnel upside down into a bowl of detergent solution. Gently blow a few bubbles into the bowl to allow a film of solution to coat the inside of the funnel. Lift out the funnel and softly blow a bubble.
- Talk with the children about how the outside edges of the soap film pulled together like a balloon around the air. The soap or detergent mixture made a more flexible skin than the water makes alone.

Food Experiences

Children put many of the water concepts to use then they cook, such as boiling water to dissolve gelatin and melting ice cubes to thicken it, and freezing fruit juice Popsicles. Cooking rice for lunch offers both water absorption and volume measurement experiences. Popping corn is an exciting edible way to illustrate a property of water: high temperature changes water to steam. Children are surprised and pleased to learn that moisture inside the kernels of corn changes to steam so fast that it pops the kernel open. The steam is very evident in a plastic domed corn popper.

Brooke is absorbed as she tests for water absorption, one drop at a time.

Thinking Games: Logical Thinking and Imagining

Invent a story mixup that includes references to water concepts. Illustrate the story with mounted pictures cut from magazines or catalogs. When the concept is referred to, let it be in the form of an obvious misstatement for the children to catch and correct. Perhaps it could be a camping story where "the children could hardly wait to cool off by skating on the lake in the summer sunshine . . . or John and Anne filled the boat with heavy rocks to help it float on the water . . . or Dad put the teakettle in the icebox to boil the water for soup."

Creative Movement

1. ***Bubbles.*** Add a soap bubble movement stimulus to your collection of emergency ideas (ideas to use when you have a group of children ready to do something *that isn't quite ready for the children*). Tell the children that you will blow imaginary bubbles to them to catch. "Here's a high one . . . catch this one on your elbow . . . your shoulder . . . your chin. Don't let this one touch the ground . . . catch this one on a fingertip and blow it back to me . . . pretend that your hands are made of soap film. Blow into them until they can't stretch any more and they pop."
2. ***Snowmen.*** Guide the children as they roll imaginary snowmen, rolling slower and slower as the ball gets larger. "Let's make a smaller ball for a chest- . . . now a smaller ball for a head . . . lift them into place. Oh! There it goes . . . snowballs this size are very heavy. Now, be the snowman yourself, all curvy and cold and tall. But wait, the sun is beginning to shine and warm the air. Oh! What's happening to your arms and your body . . . " Continue

with your suggestions and the children's responses until the snowman is a puddle that will soon evaporate and vanish without a trace.

3. Take the children on an imaginary trip to a lake in summer where they can pretend to float on their backs and stomachs while the water holds them up. They can invent ways to swim, dive, and row a boat. "Now it's winter and we can't swim, but it's so cold that we can . . . "

SECONDARY INTEGRATION

Keeping Concepts Alive

1. Discuss evaporation whenever clothes have to be dried at school—after play in the snow or a fall into a puddle. It's especially reassuring to a child who is worried about staying neat to know that an accidental stain can be washed out, the moisture will evaporate, and the garment will look fine again. When children come to school in raincoats and boots, look for drops of water still on the outside of the garments, evidence that these materials don't absorb water.
2. Examine puddles in the playground after a rain. Try to mark the outline of the puddle size with stones when it is at its fullest level. This will make it easier for the children to make comparisons from day to day as they watch for changes in the puddle. They will remind you if you should happen to forget the puddle-checking ritual. Those in the Temperate Zone might be able to see the puddle freeze and thaw as well as evaporate. Relate that cold air causes the water to freeze.

Relating New Concepts to Existing Concepts

Some of the relationships between air and water can be observed in caring for classroom fish and plant life; some can be effectively demonstrated with simple experiments.

1. Children can see evidence of *air in water* when they watch a covered jar of water that has been allowed to stand in the room. Rows of tiny air bubbles will appear on the sides of the jar. Read the story *Plenty of Fish,* by Millicent Selsam. In it, Willie learns that fish must have air to live and that water takes in air. He discovers that fish have their own way of getting the air they need from the water. Point out the safety fact that children cannot get air from the water when they swim. They must learn how to breathe, how to let the water hold them up, and how to use their arms and legs to move along when they swim.
2. Plant dependence upon water can be seen quite well in thin-leaved plants that have gone without water for a weekend. An avocado plant droops dramatically, but recuperates within an hour or so after a good watering.
3. Air pressure (see chapter 7, What Is Air?) can be used to empty the aquarium, to drain water from a large waterplay tub, or to provide outdoor waterplay

Make a siphon by completely filling approximately a yard (1 meter) of tubing with water, pinching the ends together to keep air from entering the tube. Place one end of the tube in the water and the other in the bucket below. The water will drain into the bucket, unless air entered the tube. Explain that air presses on the surface of the aquarium water, pushing water up the tube. Read about how liquids are moved from one place to another in *Messing Around With Water Pumps and Siphons*, by Bernie Zubrowski.

4. Point out the crucial role of water in ecological relationships. It is one of the nonliving substances that all living things depend on to stay alive. When too many waste materials are emptied into lakes, streams, and rivers, or when chemicals seep into underground water supplies, living things may have less water than they need. No way to make new water has been discovered. It is used over and over again in nature so people must learn to keep the world's water clean. Investigate the way water is cleaned in your area. Talk with your children about the simple ways they can keep from wasting this precious resource. Measure how much water runs from the school sinks in 1 minute, in terms of glasses of water. Talk about the need every person, animal, plant, tree, and the land itself has for water. Children who do *not* let water run needlessly are helping our planet Earth.

Family Involvement

There are many opportunities for families to point out examples of water and water/air concepts at home: moisture condensing on mirrors and windows at bath time; steamy kitchen windows when dinner is cooking and the "smoke" of moisture condensation that emits from clothes dryer vents in cold weather. All of these amplify the classroom experiences.

Parents are the best supervisors of the field trips that extend water learnings in the most concrete ways, especially the swimming pool and beach visits. There, children can stride through shallow water to test its weight and substance. With encouraging parents standing by, children can learn that the buoyancy principles work to support their own floating bodies.

EXPLORING AT HOME*

Icy Shapes. You have better access to freezing temperature in your kitchen than we have at school. Enjoy discovering with your child that water freezes in the shape of its container. For this experiment, freeze water in assorted plastic or metal containers of different sizes and shapes: thimble-size plastic tops for spray cleaner bottles, nesting measuring cups, for example. Find out together what happens to a damp mitten in the freezer. See what happens when you drape a wet paper towel over an inverted plastic bowl and freeze it. Then remove the

*These suggestions may be duplicated and sent home to families.

solid towel. Enjoy the beautiful results when you carefully squeeze drops of water from an eyedropper onto a piece of foil and freeze them. Do the same with a light coat of water droplets from a spray bottle.

Slippery Safety. (A winter lesson for children in northern climates.) Fill a deep, clear container with water. Put it in the freezer, or outdoors in below-freezing weather. See what's happening to it every few hours. When it has formed a layer of ice across the top, let your child look carefully to see that there is still water beneath the ice. The ice forms from the outside edges before it freezes in the center. The same is true when ponds or streams freeze outdoors. The ice might look safe enough to slide on near the edge, but it might not be thick and strong enough to hold a person near the center. Talk about your family rules about safe sliding and skating places.

RESOURCES

AGLER, LEIGH. *Involving Dissolving*. Berkeley, CA: Lawrence Hall of Science, 1990. Dissolving, evaporating, and crystal activities. Paperback.

______. *Liquid Explorations*. Berkeley, CA: Lawrence Hall of Science, 1990. See above. Explorations of liquid properties. Paperback.

BROEKEL, RAY. *Experiments With Water*. Chicago: Children's Press, 1988. Experiments with temperature changes, capillary action, surface tension, and buoyancy for children ready for additional challenges. (Photographs show children handling equipment that should be used under adult supervision.)

SPRUNG, B., FROSCHL, M., & CAMPBELL, P. *What Will Happen If . . . Young Children and the Scientific Method*. New York: Educational Equity Concepts, 1985. Includes activities for exploring liquids; water/sand flow comparisons. Paperback.

9

Weather and Seasons

"Who turns on the sun? Where does the wind come from?" Early in life children wonder and worry about the weather. Exploring simple concepts about the sun, water cycle, and moving air makes weather something interesting to observe rather than something to endure. Hearing of the constancy of the sun and the regularity of the seasons adds to children's confidence in life's predictable elements. The following concepts are presented in this chapter:

- The sun warms the Earth
- Changing air temperatures make the wind
- Evaporation and condensation cause precipitation
- Raindrops can break up sunlight
- Weather can be measured
- Earth's movement makes the seasons

In the activities that follow, children feel the sun's effects, observe the effect of warm-air movement, observe a small cloud in a cup, simulate a rainbow, record the weather, and imitate the movement of the Earth around the sun.

CONCEPT: The sun warms the Earth

1. What feels warm on a sunny day?

LEARNING OBJECTIVE: To become aware of temperature differences in direct sunlight and shade, and infer the source of the warmth.

Group Activity: Do this activity on a mild, sunny day. You will need sand, water, and four containers of equal size.

1. Early in the day encourage children to share ideas about the day's temperature. How did they decide what to wear? How else are they affected by the temperature? "Why is it warm today? Let's see if we can find out."

2. Let children spread a layer of sand in two of the containers and pour water in the other two. Let them touch the sand to decide if both pans of sand feel about the same

temperature; repeat with water. Move outdoors with the containers. Place one sand and one water container in heavy shade; place the other sand and water containers in the sun. Plan to return to these places in an hour. "Do you think both pans of sand will feel the same when we come back?" Return to check the predictions. Which feels warmer? What might have caused the difference? Where do the children feel warmer, in the sunlight or in the shady place where a building or tree blocks the sunlight? (The sun warms our Earth—the land, air, and everything the sunlight reaches.)

3. Move around the school area touching the building, blacktop, and sidewalks on the sunniest side; then repeat on the shadiest side. Which side feels warmer? Feel a car parked in the sunlight, then one parked in the shade. Feel the top of a rock on the ground; turn the rock over to feel the underside. Is it as bright and warm in the shadow of the school building as it is on the sunny side? If the children haven't noticed, comment that the sun gives the Earth both warmth and light. In places where the sun's warmth and light are blocked, it is cooler and darker. Even thick clouds can block the sun's light and warmth somewhat. "Will it be just as warm tonight as it is now in the sunlight? Find out and tell us tomorrow."

4. Return to the sand and water pans in the sunlight. Compare the temperatures. Is one warmer than the other even though both were in the sunlight? Do any children remember a visit to the beach? Which felt warmer on bare feet, the sand or the water? Compare the warmth of the water and the sand with a sunlit grassy area. Help children decide whether some things take up more warmth from the sun than others. (The land gets warmer in the sun than water does.)

Read *Come Out Shadow, Wherever You Are,* by Bernice Myers. Children will be able to spot Davey's confusion if they closely observe the sketches of the sun in this picture book. Ask when Davey would feel warmer and when he would feel cooler.

CONCEPT: Changing air temperatures make the wind

1. What makes air move as wind?

LEARNING OBJECTIVE: To observe that warm air moves upward.

MATERIALS:

Unshaded table lamp (60- or 75-watt bulb)

Strips of lightweight paper about $\frac{3}{4}$'' × 4'' (2 cm × 10 cm)

SMALL GROUP ACTIVITY:

Part 1

1. The A.A. Milne poem "Wind on the Hill" makes a nice introduction to this activity. Encourage children to tell how they know the wind is blowing, since wind is invisible. Summarize their ideas with a comment about knowing that wind is blowing because of what it does to things we *can* see. Children can find out something about how air moves as wind by experimenting with air that is heated.
2. Let children feel the air just above the unlighted light bulb. Give each a strip of paper to hold by one end just above the bulb. What happens to the other end of the paper? Repeat with the bulb turned on. (Urge caution near the hot bulb.) How does the air above the bulb feel now? What happens to the paper in the warm air? Recall

that when air is moving, it pushes against things. We can tell which way the air is moving because of the direction it pushes things. Encourage children to observe the direction the tip of the paper strip is pushed by the hot air, keeping their hands as steady as possible. (Be sure to tell children, "This is a *safe* way to experiment with heated air. We *never* put the paper on something hot. It could burn.")

Does the free end droop down or is it being pushed up so that it is straight or moving up somewhat? (Hot air rises.) Have children seen other evidence of hot air rising? Smoke (carbon particles mixed with heated air and gases) goes up chimneys; steam rises from a tea kettle on the stove; balloonists heat the air in their balloons to make the balloon rise off the ground. "When the sun heats the Earth, the air above the warm parts of the Earth heats up, too. All the air above warm land rises, just like the air above the bulb moves up. The rising air is part of the reason for wind."

Part 2

Clear plastic tumbler

Chilled water

Steaming water

Small, clear prescription vials

Food coloring

GETTING READY:

If water is heated in the classroom use a safely located electric pot. *Do not use immersion water-heating coils directly in a cup.* They are too hazardous to use around children.

1. "Another thing happens to air that causes it to move as wind. We can see something like it using water instead of air. Let's pretend that the cold water in this glass is cold air and the hot water I pour in is like hot air." Add two or three drops of food coloring to the vial of hot water to make it distinguishable. (Children can observe the effect better when they are at eye level with the tumbler.) What do children see? How is the dyed hot water moving? Watch for a few minutes without disturbing the tumbler. Repeat with the glass of cold water and a few drops of dyed cold water. What happens? Does the cold dyed water move slowly upward and stay around the top in a layer or does it mix with all the water right away?
2. Comment, "Water and air get lighter when they are heated. Cold water and cold air *move under* the heated water or heated air and push the hot air or hot water upward. We saw it happen with the colored water but we can't see the same thing happening outdoors. Hot air is lighter than cold air, so cold air rushes under hot air and pushes it up. When that happens we feel the cold air rushing under the warm air. The air moves fast. This makes wind. Air is always moving like this somewhere around the Earth. Warm air moves up and cooler air rushes in under it, pushing against leaves, flags, people, *everything*.'

3. Encourage children to find out at home what they can see when freezer doors are opened in warm, steamy kitchens. Which way does that foggy air move?

Note: The movement of suddenly heated air and fast rushing cold air is the cause of thunder. When lightning streaks through a cloud it moves so fast that it heats air around it very quickly. The cold air pushes under the hot air so fast that it makes the loud sound we call thunder. Let children listen to the very small popping noise they can make by drawing their lips over their front teeth, compressing their lips together, then pushing air out as they would to make the sound of a "b." A small popping noise is made by fast moving air from a small space. Thunder is made by huge amounts of fast-moving air.

CONCEPT: Evaporation and condensation cause precipitation

1. How is rain formed?

LEARNING OBJECTIVE: To observe water vapor condensing when it meets cold air, simulating a cloud and rain.

MATERIALS:

Foil potpie pan

Clear plastic tumbler

Dark paper or folder

Water, ice cubes

Safe water heater, or thermos of boiling water

Magnifying glass

GETTING READY:

Experiment at home. Form a backdrop with dark paper to make the "cloud" more visible.

Try to provide more than one setup to avoid crowding observers. Repeat as needed (Figure 9–1).

SMALL GROUP ACTIVITY:

1. Recall the earlier experience with air temperature changing water forms. "Let's watch what happens in this cup and do some pretending." Let children examine the cup and pan to be sure they are dry and free of holes. Caution the children about staying back while hot water is poured into the tumbler. "The cup *looks* empty; do you think it is? Is it easy to see through the cup of air now?" Let children add ice to the pan and feel the air temperature just below it. Pour hot water into the cup to the depth of 2" (5 cm). Let children feel the *air temperature* in the cup. "Let's pretend the cold air under the pan is like the cold air far above us. We'll watch for changes in the cup for a few minutes." Immediately cover the cup with the pan of ice. "What's happening?" (Evaporation is occurring: tiny bits of water are mixing with air to form water vapor.) "Is the cloud (fog) moving up to the cool air? Let's lift the pan to see what's happening underneath. What does it look like?" Look through the magnifier at the collected droplets. "How could those drops get on a dry pan?" There should be large drops falling before long. "What does this remind you of?" Help children recall water bits going into warm air; warm air rising up to the cold air; moist cooling air changing to larger and larger drops; drops falling back down as water again.
2. Repeat another day, using room temperature water. Put the setup on a warm, sunny windowsill. Does the same thing happen?

FIGURE 9–1

Note: To be accurate, condensation occurs in clouds when droplets collect on dust particles to form raindrops. We should say that the cloud in the cup is formed *almost like* the way rain is formed. Be certain that children know that air far from the Earth is cold because it is *far from Earth's warmth.* Air close to the Earth is warmed by reflected heat from the Earth. (Be sure that children understand that there are no pans of ice in the sky.)

2. Do water drops change in freezing air?

LEARNING OBJECTIVE: Children will observe water changing to ice and ice changing to water.

MATERIALS:

Medicine dropper (Narrow tips make more spherical drops)

Cookie baking pan

Aluminum foil

Water

Freezer or below-freezing weather

SMALL GROUP ACTIVITY:

1. "We watched water vapor get cold and change to drops of water. What do you think might happen to the drops if they got freezing cold before they fell? Let's find out."
2. Place foil on the baking pan. Carefully squeeze out a drop of water onto the foil for each child, spacing drops so they can't touch. Carry the pan to the freezer or to a sheltered spot outdoors on a freezing day.
3. Return to the freezer in 10 minutes. Show the frozen drops to the children. "Is this how they looked before? What happened?" Quickly spoon a frozen drop into the palm of each child's hand. "What's happening now?" Later comment that large frozen raindrops become ice (*sleet*). Frozen water vapor becomes snowflakes.

CONCEPT: Raindrops can break up sunlight

1. Can we make a rainbow?

LEARNING OBJECTIVE: To become aware that sunlight and raindrops can make a rainbow.

MATERIALS:

Shallow baking pan

White cardboard

Small pocket mirror

Small pitcher of water

A sunny location

SMALL GROUP ACTIVITY:

1. Initiate a discussion about rainbows and the joy of being lucky enough to see one, or even a part of one. Comment that rainbows form when sunlight happens to shine just the right way into air that holds just the right amount of water vapor after a rain. "We don't often see rainbows, but we can use a mirror and water to send back some sunlight through the water to make a bit of a rainbow."
2. Let children take turns holding a piece of white cardboard near a sunny tabletop or windowsill while other children take turns trying to reflect a spot of sunlight from the mirror to the cardboard. The mirror should be held at a 45° angle at one end of the baking pan. When the spot of light hits the cardboard, slowly pour water into the pan to cover about 2" (5 cm) of the mirror. Small adjustments to the angle may be needed until the white light is broken up into the spectrum ("like a bit of rainbow") on the cardboard.
3. Help children empty the water back into the pitcher to try reflecting the sunlight without the water, then with it.

Note: A deeper learning experience about light refraction can be found on page 255. It should be preceded by other learnings about light, however.

CONCEPT: Weather can be measured

1. How does a thermometer work?

LEARNING OBJECTIVE: To become aware that thermometer liquids expand or contract as temperature affects them.

MATERIALS:

Cooking thermometer

Hot water

Ice cubes

Small container

Outdoor thermometer

LARGE GROUP ACTIVITY:

1. Recall the experience of feeling warm in the sunshine and cool in the shade. "How else can we find out how warm the air is?" Show the thermometer. Record the level of the liquid column (the current indoor temperature). Ask for predictions about where the liquid column might move if the thermometer is taken outdoors for a while. Place the thermometer outside and compare the difference after a while.

2. Examine the cooking thermometer. Note that the numbers start with higher readings than the outdoor thermometer since food is cooked at hotter temperatures than the weather reaches. Put the thermometer into a container of hot water. "Is the liquid going up or down?" Record the final temperature. Ask for predictions of what will occur if ice cubes are added to the water. Let children add them. Check the results (after guiding children through the creative movement below). Read thermometers and record new temperatures. Make a plan to check the outdoor thermometer daily. (Thermometer must be in the shade.) Keep a record of your readings and the daily water forecast.

Thermometer Movement: "Find a space where you can stand without touching anyone. Pretend that you are a special liquid that gets bigger when it is warm and smaller when it is cool. Now you are going into a long, skinny tube with a round space at the bottom. Are you standing straight up with stomachs pulled in tight so that you fit into the thermometer tube? I'll tell you when the sun is warming the air around you. When you get warm and grow bigger, you can only go up in your skinny tube. When you shrink and get smaller in cold air, you can only sink straight down.

"Now the sun is shining in a clear blue sky. It is lunchtime. The air is getting hotter and hotter. You have to get bigger and bigger because the air is so warm. Oh! Your fingertips have reached all the way up to 40° on the number marks. I'll pretend to carry you indoors where it is cooler. The sun isn't shining inside. Ah, it feels cooler here. What will happen inside the tube when you are cooler? There you go shrinking down a bit. It must be about 25° in here. Now I'll take you outdoors again. It's getting warmer. But wait, a big cloud comes between me and the sun. I can't see the sun now. What happens to you? Do you get bigger or smaller? I see some liquid columns starting to shrink down when the sunlight is blocked and the air feels a bit cooler. Now it is night; the sun isn't shining on our part of the Earth. You are shrinking down more and more because it's cooler than it was in the daytime sunlight. Let's see what happens when a snowstorm happens. Look, there are icicles and snowmen. Now I see liquids shrinking down. Let the liquid slide all the way down to the floor. Now become yourselves again."

Read "What Will Little Bear Wear?" from *Little Bear,* by Else Minarik.

2. How fast is the wind blowing?

LEARNING OBJECTIVE: To become aware that wind speed can vary.

Group Activity: Move the group to the windows. Can children tell if the wind is blowing? How? What do they see? In a group discussion, introduce the ideas that the wind, the heat of the sun, and water work together to make the weather. Winds move rain clouds and dry air.

Weather is more interesting to observe when we know many ways to describe and compare it. For generations people have judged the wind's speed by observing its effect on things around them. The following chart is part of the Beaufort scale of wind velocity and provides visual guidelines for determining the wind's speed.

What to Look For	Description	Miles per Hour
Leaves don't move. Smoke rises straight up from chimneys.	Calm	Less than 1 mph
Light flags blow; leaves and twigs move constantly.	Gentle Breeze	8–12 mph
Large branches sway. It's hard to hold an open umbrella.	Strong Breeze	25–31 mph
Whole, large trees move. It is hard to walk against the wind.	High Wind	32–38 mph
Whole trees uprooted.	Full Gale	55–63 mph

End the group discussion with the measures we take to stay safe in storms. Assure children that weather forecasters collect information very carefully so we can be warned about serious storms and find safety when necessary.

Plan to make a daily check at the windows to decide how fast the wind is moving. Record the description the children decide on in a wind chart. Children can check the direction of the wind each day by observing the school flag or by holding up a wet finger to see which side is cooled by the wind. This wind is named for the direction it blows *from.* (A north wind pushes the flag south.)

3. Can we tell how much rain falls?

LEARNING OBJECTIVE: To become aware that amounts of rainfall vary.

MATERIALS:

Widemouthed glass or clear plastic jar

Heavy rubber bands

Small plastic ruler

LARGE GROUP ACTIVITY:

1. Elicit children's ideas about why rain is needed. Discuss how every living thing needs water to survive, and also that lakes, rivers, reservoirs, and underground water stores are filled by rain. Explore the reasons that it is important to many people to know how much rain falls. (Farm children, those living in flood areas, or those living in cities dependent upon reservoirs for water supply may have information to share about rainfall amounts.)

   "Weather forecasters use special measuring equipment to find out how much rain has fallen. We can make something like it. We can put our rain gauge outside in a safe, open place. After each rain we can check to see how much rain fell." (This will be a relative measurement. It will allow comparisons of the amount of rain collected in the jar from day to day.)
2. Attach the ruler to the jar with rubber bands. Locate a safe, open place where the jar will be undisturbed. Check the level of water in the jar as soon as possible after a rain. Empty the jar. Keep a record of the number of inches (centimeters) of rain that has fallen for several weeks in your rainy season.

CONCEPT: Earth's movement makes the seasons

1. What makes day and night?

LEARNING OBJECTIVE: To become aware that night and day occur as Earth blocks and receives the sun's light.

Large Group Activity: For this activity you will need a globe on a stand, a bit of clay, and a flashlight.

1. Recall experiencing the sun's warmth and light outdoors. "When we stood in the shade it was darker and cooler. Do you know that the whole Earth does the same thing? That is the way day and night happen. Let's pretend that this light is the shining sun. This globe is a model of our huge Earth. I'll put a bit of clay here on the globe to show where our part of the world is. Watch the clay as I turn the globe the way the Earth always slowly turns. Is the clay in the light now? (Turn the globe.) Now? Our part of the Earth is in a large shade (shadow) now. It's the shade (shadow) of the other side of the Earth. The sunlight is blocked, so it is dark here." Continue to turn the globe so that children can see the marked part of the globe alternately in the light and in the shade. "What do we call the time when our part of the world is in the Earth's shade? When it is in the sunlight?"

2. Take the group to the playground or gym. Ask them to form a tight, round group, with all children facing outward, in the center of the space. They represent the large, hot sun that warms and lights the Earth. You represent the much smaller Earth moving around the sun in space. Walk slowly around the "sun" counterclockwise, keeping your "orbit" as far from the "sun" as possible. "Our Earth is always traveling around the sun. It never stops. The Earth and sun are so far from each other in space that it takes exactly one year for the Earth to make its trip around the sun."

3. When you complete "one year's journey," add that the Earth also spins around as it moves around the sun. "Our Earth is so big that it takes a whole day and a whole night to turn around completely." Start another trip around the "sun," this time spinning as you travel. "Pretend that my face is one side of the Earth and my back, the opposite side of the Earth. When I am facing you as I turn, you are shining on me, and it is daytime on my part of the Earth. When I turn away from you, you shine on my back. My body blocks your light, so it is dark nighttime for my part of the Earth." Move and turn a few times saying, "Now it is daytime for my part of the Earth, now it is nighttime." Let children take turns being the spinning Earth traveling around the sun, identifying facing the "sun" as daylight, turning away from the "sun" as night.

Read *The Day We Saw The Sun Come Up,* by Alice Goudey.

2. How do the shining sun and moving Earth make seasons?

LEARNING OBJECTIVE: To become aware that seasons change as the tilted Earth moves around the sun.

Large Group Activity:

1. Join the children who should be seated with legs crossed in a circle on the floor. "Our Earth doesn't travel around the sun as we were turning around our pretend sun. We were standing up straight. The Earth is tipped to one side (tilted) as it spins." Point to the slant of the globe support intersecting the globe from north to south. Touch the North and South poles. "We can think of these as the topside and underside of the Earth. If there were a giant stick running from the North Pole to the South Pole, it would be slanted, just like this. That slant makes a difference in our lives."

2. "Is something different outside when you go to bed in winter than it is when you go to bed in summer?" Help children distinguish between the long days of summer when it may still be light outdoors at bedtime, and the shorter days of winter, when it may even be dark at dinnertime. "What else is different about the outdoors in summer and win-

ter?" Summarize by saying that the sun doesn't warm the Earth as much in winter as it does in summer.

3. "Let's find out how that happens." Ask one child to sit in the center of the circle, shining the flashlight on the globe to represent the sun. The circle of children will help the model Earth make its year-long trip around the "sun." Seat yourself in the east quadrant of the circle. Arrange the globe so that the North Pole is pointed toward yourself and *away* from the "sun." "Let's start the Earth's journey here. The clay showing our part of the Earth is here. It is farther from the sun than the south underside at this time in the journey. Our part of the Earth gets less sunlight. Daylight ends before dinnertime. The ground and air are cold. It is wintertime." And that, "winter begins on the day that has the fewest hours of daylight, about December 21." That day is the shortest day of the year because our part of the Earth is as far from the sun as it will be all year. The shortest day and the longest night we have each year marks the beginning of winter."

4. Slide the globe to the child at your right. Show the children how to slide the globe carefully from person to person so that it tilts in the same direction throughout its whole trip around the sun. Each child could slowly turn the globe one full revolution so that your part of the Earth is in the path of the flashlight beam and in the shadow of the beam (representing day and night). As the globe moves toward the children in the north quadrant of the circle, comment that slowly, day by day, our part of the Earth gets more light and warmth. The days get a little bit longer because we have more sunlight. When the globe reaches the child in the northernmost part of the circle, say, "Now our part of the Earth is close enough to the sun so that the daylight lasts just as long as the night. That day is the beginning of spring, about March 21."

As the globe is carefully moved toward the west side of the circle, ask children to notice whether the North Pole or the South Pole is away from the sun. "Is our part of the Earth closer to the sun now? What time of the year does that happen? Add that when our part of the Earth is as close to the sun as it will get, daylight lasts until bedtime. It is summertime. Nights have the fewest hours of darkness then. Summer begins on the longest day of the year, about June 21."

"Now our part of the Earth is slowly moving away from the sun. The days slowly get shorter; the nights get longer. The air begins to feel cooler. What time of the year does that happen?" When the globe reaches the southernmost point in the circle, comment, "Now neither the South Pole nor the North Pole is closer to the sun. In our part of the Earth, the day is just as long as the night and fall begins (about September 23). In the fall or autumn, days become shorter and nights become longer. We see the sun less. Does the sun warm the Earth as much now as it did when our part of the Earth was closer to the sun?"

5. Take time to move the globe again, but this time let the children tell about the changes in warmth and light that take place as our part of the Earth moves farther, then closer, to the sun. They could add things they like to do seasonally. Guide the children as they become the sun clustered in the center of the circle and you move your tilted body around them, as illustrated in Figure 9–2.

6. Demonstrate the "slanted Earth" posture (Figure 9–2, left) and the pattern of walking forward on one curve of the circle, turning your feet and walking backward down the other curve of the circle; turning your feet and walking forward to complete the circle (Figure 9–2, right). Maintain the tilted Earth position with right arm slanted up and out; the left arm slanting down and out. "My right hand will be like the North Pole and my left hand will be like the South Pole as I move around the sun." Let the children take turns being the slanted Earth.

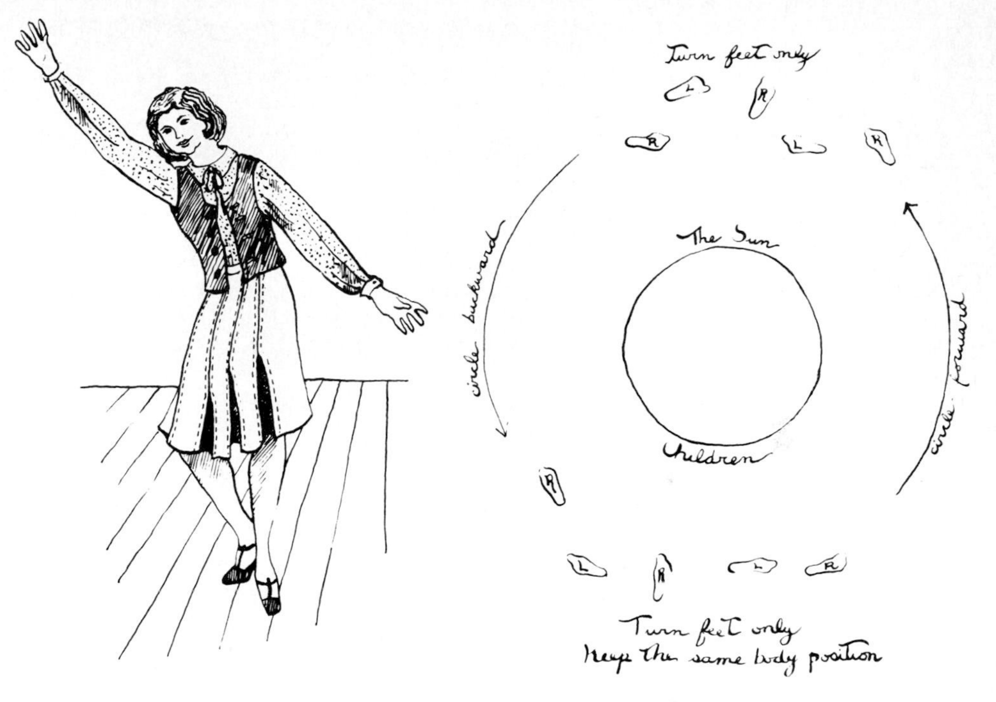

FIGURE 9–2

PRIMARY INTEGRATING ACTIVITIES

Music (Resources in Appendix 1)

1. Sing the songs that children have been singing about the weather for generations: "One Misty Moisty Morning," "Rain, Rain, Go Away," and "The North Wind Doth Blow."
2. Sing "The Eency Weency Spider" remembering to change "dried up all the rain" to "evaporated the rain." Check the rain spouts around your building after a rain.
3. Listen to "Where Does the Sun Go At Night?" sung by Tom Glazer on *Songs to Learn By.*
4. Hang a melodic wind chime near or in your classroom on a breezy day. Let the weather provide the music.

Math Experiences

Chart the weather for the remaining months of the school year.

For Younger Children. Mark off a calendar on a large sheet of newsprint. Make a yellow sun symbol in each square to remind children that the sun is shining whether we can see it or not. (Speak accurately when asking children

A bar graph produces weather records for the group to recall and compare.

about the day's weather: "Can we see the sun today?" not, "Has the sun come out today?") Cut squares of white tissue paper to tape over the sun on foggy days and opaque white cloud shapes to cover the sun on rainy and snowy days, according to the children's decision about the day's weather.

For Older Children. Record the weather on the calendar by charting the temperature, the class estimate of the wind speed, and sky conditions (clear, cloudy, rainy, sunny). At the end of the month, summarize the total number of days of each condition. A bar graph makes comparisons of the month's weather easy, providing ready answers to questions such as, "Which kind of weather did we have the most often?" and, "How many more days were cloudy than sunny?" Go back to compare monthly or seasonal summaries.

Stories and Resources for Children

BAUER, CAROLINE (Ed.) *Windy Day.* New York: J.B. Lippincott, 1988. This anthology of stories and poems is devoted to windy weather.

BRANLEY, FRANKLYN. *Flash, Crash, Rumble and Roll.* New York: Thomas Y. Crowell, 1975.

______ . *Hurricane Watch.* New York: Thomas Y. Crowell, 1985. Facts are presented about the nature and power of hurricanes, and the protective measures that keep people safe during the storm.*

______ . *Snow Is Falling.* New York: Harper & Row, 1986. The very readable text describes both the benefits and hazards of snow. A temperature experiment is included. Paperback.

______ . *Sunshine Makes the Seasons.* New York: Harper & Row, 1988. A good followup for the seasons activity. Illustrations clarify how Earth's movements alter the amount of sunlight for different areas.*

______ . *The Sun: Our Nearest Star.* New York: Harper & Row, 1988. Shows a home heated by solar energy. Paperback.

______ . *Tornado Alert.* New York: Harper & Row, 1990. The nature and incredible power of tornadoes will intrigue children. The safety tips will reassure them. Paperback.*

CALHOUN, MARY. *Hot-Air Henry.* New York: William Morrow, 1981. Henry the cat has unexpected adventures in a hot-air balloon. Exciting illustrations by Erick Ingraham clearly show how the balloon's movement is controlled. Paperback.

DEPAOLA, TOMIE. *The Cloud Book.* New York: Scholastic Book Services, 1974. "If you could hop on a bird and fly way up, you would see . . ." Factual information about cloud types follows. Amusing illustrations. Paperback.

DORROS, ARTHUR. *Feel the Wind.* New York: Thomas Y. Crowell, 1989. Clear, simple and accurate illustrations in a charming text. Instructions for making an easy weather vane are included.

EMIL, JANE. *All About Rivers.* Mahwah, NJ: Troll, 1984. Includes brief descriptions of floods and flood control. Paperback.

ENGDAHL, SYLVIA. *Our World Is Earth.* New York: Atheneum, 1979. The text includes an imaginary description of Earth's climate to a person from another planet. Children will appreciate Earth's blue sky and the warm sunlight when they complete this fantasy exploration.

GIBBONS, GAIL. *Sun Up, Sun Down.* New York: Harcourt, Brace, Jovanovich. 1983. This story follows a young child through a day, aware of the sun's effects. The water cycle is included.

GIBBONS, GAIL. *Weather Words and What They Mean.* New York: Holiday House, 1990. Clearly presented weather definitions.*

GOUDEY, ALICE. *The Day We Saw the Sun Come Up.* New York: Charles Scribner's Sons, 1961. A beautifully written and illustrated story that explains the occurrence of day and night at the young child's level.

KAUFMAN, JOE. *Joe Kaufman's Big Book About Earth and Space.* New York: Golden Books, 1987. This book compiles capsules of child-level information about the geology, geography and weather of our planet Earth. A good classroom reference with basic answers to children's questions about scary events: earthquakes, volcanoes, tornadoes, and hurricanes and other storms.

MARCUS, ELIZABETH. *Our Wonderful Seasons.* Mahwah, NJ: Troll, 1985. Simple experiments are described that could be carried out independently as enrichment activities by primary-grade children. Paperback.

MINARIK, ELSE. *Little Bear.* New York: Harper & Row, 1957. Chapter 1 provides a good take off for discussing adaptations to cold weather for young children.

MONCURE, JANE. *What Causes It? A Beginning Book About Weather.* Elgin, IL: The Child's World, 1977. Simple explanations of rain formation, thunder, and rainbows. Attractively illustrated. Good for young children.

MYERS, BERNICE. *Come Out Shadow, Wherever You Are!* New York: Scholastic Book Services, 1970. Children will find the humor in this story if they carefully watch the sketches of the sun and cloud positions.

*Starred references, written at the young child's level of understanding, can help teachers with minimal backgrounds in science expand their knowledge base.

POLLACO, PATRICIA. *Thunder Cake.* New York: Philomel Books, 1990. A wise grandmother dispels a child's fear of thunder as they bake together before a storm breaks. Appealingly illustrated.
ROCKWELL, ANNE. *My Spring Robin.* New York: Macmillan, 1989. The simple text extends the concept of seasonal change for preschoolers.
SABIN, FRANCENE. *The Seasons.* Mahwah, NJ: Troll, 1985. A good reference on the seasonal changes taking place as Earth orbits the sun. Paperback.*
SABIN, LOUIS. *Weather.* Mahwah, NJ: Troll, 1985. Basic information about weather conditions and forecasting make this a good reference for primary-grade classrooms. Paperback.*
SANTREY, LAURENCE. *What Makes the Wind?* Mahwah, NJ: Troll, 1982. This simple, clear book can be read selectively to young preschoolers, and in its entirety to older children. The breezes and gales illustrated in Bert Dodson's watercolors almost lift off the pages. Paperback.*
SCARRY, RICHARD. *Great Big Air Book.* New York: Random House, 1971. In the story "A Spring Day," Miss Honey shows how rising hot air creates a small wind in her kitchen. Other weather stories are included.
SCHNEIDER, HERMAN & NINA. *Science Fun with a Flashlight.* New York: McGraw-Hill, 1975. Sunbeams, variations in sunlight, and rainbow formation are explained. Attractive, lighthearted illustrations.
THOMPSON, BARBARA, & OVERBECK, CYNTHIA. *The Winds That Blow.* Minneapolis: Lerner, 1977. The simple text and attractive illustrations will sharpen children's awareness of differing velocities of air movement.
TRESSALT, ALVIN. *Hide and Seek Fog.* New York: Lothrop, Lee & Shepard, 1965.
WYLER, ROSE. *Raindrops and Rainbows.* Englewood Cliffs, NJ: Simon & Shuster, 1989. Simple text and interesting experiments, such as measuring the size of raindrops, exploring aspects of rain, clouds, and rainbows.
ZOLOTOW, CHARLOTTE. *The Storm Book.* New York: Harper & Row, 1952. This classic book has helped many children master fears of storms, but inaccuracies about thunder and lightning should be corrected as the story is read.

Poems (Resources in Appendix 1)

A good poem for stimulating weather imagery is "Spring Morning," by A.A. Milne, from *When We Were Very Young.* The following poems also offer excellent weather imagery.

From A.A. Milne's *Now We Are Six:*
"Waiting at the Window"—Which rain drop on the window will be the winning one?
"Wind on the Hill"—Reveals a child's problem-solving logic about where the wind blows, and the unsolved question of where the wind comes from.
From Robert Louis Stevenson's *A Child's Garden of Verses:*
"The Wind"—Another child tries to understand the wind.
"Rain"—Four classic lines of verse.
"Bed in Summer"—A good follow-up for the seasons activity.
From *Poems Children Will Sit Still For:*
"Weather," by Eve Merriam.

Art Activities

Weather Mobile. Let children help you create weather symbols to hang in your classroom. Cut out pairs of cloud shapes from newsprint. Staple them together over wire coat hanger. Let children stuff them with shredded paper or polyester fiber to create a soft sculpture appearance. Place ends of plastic silver tinsel on strips of tape. Fold the tape lengthwise over the bottom of a white wire coat hanger to simulate rain. Let children crayon spectrum colors on a cardboard rainbow arch. Add a yellow construction paper sun to the mobile.

Fog Pictures. Let children draw outdoor scenes on light blue paper. Attach a sheet of white tissue paper of matching size to the top edge to give the effect of fog. The "fog" can be lifted by folding back the tissue paper.

Rain Painting. See what changes will be created by holding children's finger paintings outdoors in a light rain for a few seconds.

Sidewalk Painting. Give children wide brushes and cans of water to paint huge designs on the sidewalk on a cloudy day. Ask for predictions about what will happen to the designs when the sun comes out or a warm wind blows over them.

Food Experiences

For centuries warm air has been used to preserve fruits, vegetables, fish, meat, and herbs. Children enjoy preparing dried apple rings. Let them wash several apples. Then you core the apples and slice them crosswise into $\frac{1}{4}$" (0.6 cm) rings. Children then drop the apple rings into a bowl of water to which 2 tablespoons of lemon juice have been added. Pat rings dry. String them on a long, clean dowel or on a heavy cord. Tie the cord or place the dowel across a sunny, warm corner of the classroom to dry for several days. Apples will be dry to the touch, darker in color, and slightly rubbery. Rinse slices, dry, and eat.

Experiment with drying small clusters of grapes on a towel-covered tray placed in the sun. Turn grapes each day. Cut fresh corn from the cob. Dry in the same way. Feed the dried corn to the birds, or soak it overnight, simmer for 15 minutes, and serve. Bring samples of dried legumes, dried fruits, or dried beef, if possible.

Thinking Games: Imagining

Print as a story, or tape-record children's responses to stimulus phrases such as "If I could drift on the wind like a bird . . .," "If I were a raindrop . . .," "If I could sail away on a cloud . . ." Read the completed story or play the finished tape for the whole class to enjoy. Make a booklet of the ideas evoked by the children, including illustrations by each idea's author.

Field Trips

After a rain stops, tour the school neighborhood to examine the effects of wind and rain on the environment. Where has water collected in a puddle? Why there? Is a wooden sandbox seat as wet as a blacktop area? Is the ground under the playground swings as dry as a graveled area? Have leaves been blown down (small branches, large branches)? What might these signs indicate about the speed of the wind? Was there enough rain to flood the gutters or to wash earthworms out of their tunnels?

SECONDARY INTEGRATION

Keeping Concepts Alive

1. Expand the weather chart with questions about adaptations children may have made to cope with the day's weather. It can become part of the day's routine to inquire about how the weather affected their lives positively or negatively.
2. If a small dehumidifier is available, examine it on a humid day with the children. See how this adaptation to change a room's climate uses the process of condensing water vapor to water. Unless your dehumidifier poses a special safety hazard, let the children feel the condenser coil before turning on the unit. Switch on the unit and let children hold a piece of paper next to the condenser, then in front of the blower screen, to discover which way air is being moved. Watch the changes gradually taking place on the condenser coils. Does the coil feel the same after the unit has run for a while as it did before? Let children help empty the water from the collection pan. Help them relate this process to the cloud in a jar and to clouds in the cold atmosphere.

Relating New Concepts to Existing Concepts

1. A miniature water cycle can be observed in a classroom terrarium. At times condensation will be visible inside the glass enclosure. At other times it will appear to be dry. The plants survive without additional watering for many weeks. Why?
2. Recall the experiences with water changing forms; with air taking up water; with the effects of moving air; with water pushing up objects that are less heavy than water; and with air supporting gliders.
3. Note that the water cycle (the evaporation/condensation process) is one of nature's ways of keeping our limited supply of water clean.

Family Involvement

Encourage families to show children the cold, wet air that rolls down from an open freezer door on a humid day, or the direction hot steam takes when it escapes from pots of simmering food or from a teakettle. Suggest that families owning Swedish holiday candle chimes examine the before and after effects of heating the air below the figures with the small candle flames. Note how the blades of the candelabra are bent to catch the rush of hot air and begin to move. Tell families where nearby homes or businesses have visible solar energy panels on their rooftops. They could mention that the sun's energy can be used to heat water to those homes. Mention the location of windmills in use locally to generate electricity. Suggest that families discuss the things they do to save energy used to heat and light homes.

EXPLORING AT HOME*

Ups and Downs of Air. Find out together if there is moving air in your house or apartment: tape a small strip of thin paper to one end of a piece of thread. Tape the other end of the thread to a doorframe to let the paper dangle loosely. Quietly watch with your child as the paper drifts on the air current. Next, explore the ups and downs of air movement. *Carefully,* let your child hold his or her hand just above a lighted 100-watt light bulb to feel the warm air. Then, hold a $\frac{1}{2}''$ strip of tissue paper by one end, horizontally above the bulb. Watch the free end of the paper strip flutter up as the warm air rising from the glowing bulb pushes the paper up. Next, let your child hold the strip of paper just below the freezer or refrigerator door as you open it just a crack. The cold air pushes the paper end down. Talk with your child about how wind happens: all over the world some air is being warmed by the sun, and some air is cool. The cool air moves down under the warm air as the warm air rises. All that pushing and rushing air is what we call wind.

Dry Facts. Join your child in noticing and pointing out common, everyday ways that air picks up moisture: a withered carrot or apple from the back of the refrigerator; a piece of dried-out bread; once-damp mittens, now dry; air-dried dishes on the kitchen counter . . . all are examples of how air takes moisture from things it touches.

Getting It Together. Point out to your child how moisture from warmed air condenses into droplets when it meets a cooler surface: under the lid of a carry-out cup of hot beverage; on the bathroom mirror after someone's warm shower.

Freeze some water in a wide and deep plastic bowl. Check the freezing process every hour to notice how the water begins to freeze from the edges of the bowl long before the middle freezes. Mention that this also happens outdoors when ponds or streams freeze. This is important safety information for children living in cold climates. Ice may be thin and unsafe to walk on in the middle of a pond, even when it is thick and safe to walk on near the shore.

RESOURCES

BROEKEL, RAY. *Storms.* Chicago: Children's Press, 1982. Excellent photographs of storms.

EARTHWORKS GROUP. *50 Simple Things Kids Can Do to Save the Earth.* Kansas City: Andrews & McMeel, 1990. Two weather-related eco-experiments are described, dealing with the effects of smog and acid rain.

*These suggestions may be duplicated and sent home to families.

FORTE, IMOGENE, & FRANK, MARJORIE. *Puddles and Wings and Grapevine Swings.* Nashville, TN: Incentive, 1982. Art projects and games to celebrate seasonal changes are included in this lively book of nature crafts.

LEHR, PAUL, BURNETT, R., WILL, & ZIM, HERBERT. *Weather.* New York: Golden Books, 1987. Pocket-size guidebook. Paperback.

ZUBROWSKI, BERNIE. *Messing Around with Water Pumps and Siphons.* Boston: Little, Brown, 1981. Directions given for making an inexpensive air thermometer.

Teaching Resource

Children's Television Workshop. The *BIG BIRD Get Ready for Hurricanes* kit. Hurricane safety information, planning for a hurricane "watch" and "warning." Includes booklet, game, and audiotape of Big Bird's song. $2.25 for a single copy. Mail check payable to: Children's Television Workshop, Dept. NH, One Lincoln Plaza, New York, NY 10023.

10

Rocks and Minerals

". . . Oooooh! This one has shiny speckles! Look! This one is all little rocks joined into one. These three are a family. This one has a stripe all around and around." Children relish sifting through piles of rocks to find favorites. They are awed to hear that our Earth is a giant ball of rock and that rocks can be millions of years old. Learning about the importance of rocks seems to promote feelings of security in youngsters, just as the Rock of Gibraltar symbolizes trust and stability to adults. Experiences in this chapter explore the following concepts:

- There are many kinds of rocks
- Rocks slowly change by wearing away
- Crumbled rocks and dead plants make soil
- Old plants and animals left prints in rocks
- Minerals form crystals

An unstructured experience in washing and examining ordinary highway gravel is suggested as a beginning activity. This is followed by classifying and hardness testing. Basic information is given about rock formation. Suggested experiences include grinding soft rocks, breaking rocks open to compare the fresh surfaces with the worn exteriors, pulverizing rock to compare it with complete soil, and forming crystals.

Introduction. Tuck a small rock into your hand, then comment quietly, "I have something quite small but very old in my hand. It is so old that it might have been here long before any people lived on Earth or even before the dinosaurs were alive. Can you guess what it could be?" Slowly open your hand to reveal the humble, but now significant, rock.

CONCEPT: There are many kinds of rocks

1. How do rocks look when dry and wet?

LEARNING OBJECTIVE: To notice and describe the appearance of a variety of rocks.

MATERIALS:

Bucketful of #67 washed gravel (builder's supply store)*

Dishpans, water

Few drops of detergent

Old newspapers (for under pans)

Trays for clean rocks

Cleanup sponge

Smocks for children

GETTING READY:

Fill dishpans one-fourth full with water.

SMALL GROUP ACTIVITY:

1. Invite the children to enjoy looking at and washing the rocks. "Do wet rocks look the same as they did when dry?" Listen to children's comments.
2. Leave the gravel out for children's independent examination during unstructured times, if possible.

Note: Children seem to be drawn to a material that is available in quantity for them to explore freely. A heaping dishpan of common rocks will have more appeal than a box of neatly labeled special rocks.

2. Can we find rocks that are alike in some ways?

LEARNING OBJECTIVE: To group rocks into simple classes.

MATERIALS:

Bucketful of #67 washed gravel*

Sorting containers (egg cartons, margarine tubs, divided plastic snack trays, old muffin tins)

SMALL GROUP ACTIVITY:

1. Suggest putting rocks that are alike in some way together in containers.
2. Classifying decisions are easy for many children to make (by color, size, shape, texture, pattern). If children can't get started on their own, offer easy ideas such as, "The flat rocks go here; rocks that aren't flat go there."

*#67 washed gravel consists of varied rocks (from old riverbeds) that range from almond to egg size. For school use, some building suppliers might give you a bucketful without charge. It is usually sold by the ton.

3. Which rocks are hard; which are soft?

LEARNING OBJECTIVE: To use hardness tests as the basis for grouping rocks.

MATERIALS:

#67 washed gravel

Blackboard chalk (pressed gypsum)

Pumice (from the drugstore)

Pennies

Three trays or boxes

GETTING READY:

Make two labels: *soft* and *hard*.

Place trays next to each other. Put *soft* and *hard* labels in first and third boxes; leave middle box unlabeled.

SMALL GROUP ACTIVITY:

1. "Some rocks are soft enough for a fingernail to scratch them. Rocks just a bit harder can't be scratched that way, but a penny edge will scratch them."
2. Let children experiment with scratch tests. Suggest sorting rocks by degree of hardness: put rocks scratched by your fingernail into tray marked *soft;* put rocks scratched only by the penny into unmarked tray; put rocks that can't be scratched by a penny into the tray marked *hard*.

If children find it hard to form three classes of rocks, simplify the experience to form two classifications of hardness: hard rocks, not scratched by a penny; and softer rocks, scratched by a penny. Make a separate activity to find soft rocks, scratched by a fingernail, and harder rocks, not scratched by a fingernail. Perhaps a teacher is wearing the hardest rock of all on her finger—a diamond (crystallized pure carbon).

ROCK FORMATION

As children engage in the suggested experiences, they can be fed bits of simple information about the mineral content of rocks and about the way rocks were (and are) formed.

Mineral Content

When children question the specks and streaks of color they see as they sort rocks, tell them that most rocks are mixtures of many kinds of material. The materials are called *minerals*. There are about 2,000 different minerals, but only about 20 minerals are abundant. Different mixtures of minerals make different kinds of rocks. (You might want to recall baking experiences with them. Talk about how sometimes the same materials are put together in different ways to make different things to eat. Corn bread and cookies are almost alike in ingredients.) Sometimes the mineral mixtures are easy to see as specks, sparkles, and stripes. Rocks may contain many different combinations of minerals.

Three Types of Rock Formation

Igneous. Some rocks were once mixtures that melted deep inside the Earth. When they cooled, they formed rocks.

Sedimentary. Some rocks were formed in layers or parts, like a sandwich. Layers of sand, clay, and gravel were pressed together very hard at the bottom of old lakes, rivers, and oceans. Sometimes animal shells were mixed into the rock. It takes thousands of years of very hard pressure to form rocks this way. Some of our rocks have stripes of color or line marks in them. Some feel sandy. Some have tiny rocks stuck together. They were all made under pressure.

Metamorphic. Sometimes rocks that were already formed by pressure or by the cooling of melted mixtures were changed again by more heat and pressure, changing the rocks into still different types.

Try to find out something about the rocks in your area, both the visible outcroppings and the rocks beneath the soil. Were rocks heaved up into mountains and hills in your area millions of years ago? Did ancient rivers wear away rocks and carve out valleys? Mention this to the children when a suitable time arises.

CONCEPT: Rocks slowly change by wearing away

1. Can we wear away bits of rock?

LEARNING OBJECTIVE: To observe and experience one of the ways rocks change.

MATERIALS:

Pieces of soft, crumbly rock (shale, soft sandstone)

Tin cans with lids (coffee cans, large tobacco cans, cocoa tins)

SMALL GROUP ACTIVITY:

1. Put several soft rocks in each can. Ask children to feel the rocks. Cover the cans tightly. Let children hold them tightly and shake them as long and as hard as they can. Open cans. "Has anything changed? Is there dusty stuff at the bottom?"
2. If this must be an indoor activity, try briskly rubbing two crumbly rocks together over a piece of white paper.

Group Discussion: Talk about the results of the rock-shaking activity in terms of wearing away pieces of rock. Together, look at the edges of the smoothest pebble you can find and at the edges of a freshly broken rock (one you have broken open with a hammer). Encourage children who have visited natural beaches to recall whether they found smooth pebbles or jagged rocks on the sand. Talk of how pieces of rock break from large rocks, how waves tumble the rocks against each other and smooth them, and how sand bits are rubbed off. If there are nearby rock formations with which children are familiar, speak of how they were slowly changed and worn by winds or running streams of water.

2. Can we draw with rocks?

LEARNING OBJECTIVE: To experience the idea that bits of rock wear away.

Group Activity: If there is a safe sidewalk area where this activity can be performed, give children assorted soft rocks to use for outdoor sidewalk art. Mention that cavemen made pictures on rock walls with soft rocks of different colors.

Later show the children two other rocks we use for writing: pencil "lead" (graphite) and blackboard chalk (pressed gypsum). They make marks because they are so soft that worn-away bits are left as the drawing or writing we see.

Does your schoolroom have an old-fashioned slate blackboard to draw on? (Slate is a metamorphic rock.) Perhaps you can buy an inexpensive one at the variety store. Slabs of slate can sometimes be found in salvage stores after wrecking crews have taken off old slate roofs. (They are fine to use under flowerpots on the windowsill after the writing experience is over.)

3. Do rocks look differently on the inside?

LEARNING OBJECTIVE: To discover firsthand that weathering changes the outside appearance of rocks.

MATERIALS:

Rocks

Hammer

Canvas or burlap (precaution against flying rock chips)

SMALL GROUP ACTIVITY:

1. Cover a rock with the cloth. Strike it hard with the hammer. (Children experienced with hammers can do this successfully. A two-handed grip may help.) Very hard round rocks may not split well. Other rocks will split with one hammer blow.
2. Examine the inside; compare it with the outside appearance.

Note: This is a very exciting activity that may lead to "rock fever." Children will be curious to discover what is inside a rock: the glint of mica flakes, a streak of mineral, a sleek surface. Excite children's curiosity further by stating that the worn outside surface of the rock was once the same color and texture as the inside. Ask, "What might have happened to change the outside of these rocks?"

CONCEPT: Crumbled rocks and dead plants make soil

1. What happens when we pound soft rocks?

LEARNING OBJECTIVE: To become aware that rocks are part of soil formation.

MATERIALS:

Pieces of crumbly rock (shale, soft sandstone, etc.)

Heavy paper bags

Hammers

Newspapers

Sieve

Empty can

SMALL GROUP ACTIVITY:

1. Put one bag inside another. Put a few rocks in the bags and place on sidewalk. Let children take turns pounding rocks, checking results often.
2. Spread newspaper on the ground. Fit sieve over the can. Shake contents of bags into sieve, catching spills on newspaper.
3. Keep pulverizing until bags are tattered and children are tired.
4. Examine the powdered rock in the can and save for the next experience.

2. What does soil look like?

LEARNING OBJECTIVE: To observe that many materials make up fertile soil.

MATERIALS:

Trowel

Container

Newspapers

Topsoil

Sieve, sticks

Magnifying glasses

Container of pulverized rock from previous experience

2 paper cups

Water

SMALL GROUP ACTIVITY:

1. Let children scoop up a trowelful of soil in a permitted digging area to bring indoors.
2. Spread the soil on newspapers; examine with magnifying glass. Compare with rock pulverized by children.
3. Put soil through sieve. Good topsoil contains bits of leaves, twigs, roots, and worms. Some of this matter will be left in the sieve.
4. Put some sifted soil and powdered rock in separate cups. Stir a bit of water into each. Compare.

Note: Clay is like the moist powdered rock. It must have lots of old vegetable and animal matter (and perhaps sand) added to it to make good soil for growing things.

CONCEPT: Old plants and animals left prints in rocks

Recall splitting rocks open to find unexpected color, sparkle, and texture inside. Add that prints of animals, shells, and plants that lived on Earth millions of years before people did can be found pressed between layers of rocks. These rocks are called *fossils.* (Perhaps there are some fossils in the bucket of highway gravel that the children have been using.)

Children can have fun making prints of their hands, shells, or leaves using the kind of powdered stone that became clay. They can make prints similar to fossils, though they won't be as old or as hard as fossils, of course. Related references to prehistoric animals are listed in Appendix 2.

1. Can we make a "fossil" print?

LEARNING OBJECTIVE: To become aware of one way that fossils are formed.

Let children roll out moist clay slabs about $\frac{1}{4}$'' (1 cm) thick with small rolling pins. Show them how to press a leaf, shell, or their hand firmly into the clay. It may take several tries to get a clear print. "Erase" unsatisfactory leaf prints by rubbing the clay with water and lightly rerolling. Cut the slab into a plaque shape using an empty coffee can as a cutter. Poke a hole through the top of the slab for a hanging loop of string. Scratch the child's initials in the clay. Allow several drying days. Teacher may shellac the dried clay to preserve it, if desired.

CONCEPT: Minerals form crystals

Introduction: (Do this after children understand evaporation.) Look together at some mineral crystals: a diamond or other precious stone jewelry, the rocks you cracked open to reveal glittering bits of mica or quartz crystals, or some large or small salt crystals. Say that when minerals are dissolved in liquid which then evaporates, or when minerals

Rusty puts real effort into discovering what his rock will look like inside.

melt and cool slowly, they form *crystals*. Each type of mineral forms into its own special crystal pattern. "We can try to dissolve a mineral in liquid that will evaporate to leave pretty, powdery crystals."

1. Will crystals form on coal?

LEARNING OBJECTIVE: To observe the formation of crystals.

MATERIALS:

Coal, broken into small chunks (charcoal briquettes may be used)

Glass pie pan or low ceramic bowl

Old tablespoon

Pint jar for mixing

4 tablespoons *plain* salt

4 tablespoons water

2 tablespoons *clear* household ammonia*

4 tablespoons laundry bluing

Food coloring in squeeze bottles

SMALL GROUP ACTIVITY:

1. Heap chunks of coal in pie pan.
2. Let children take turns measuring salt and water.
3. Tell children to stand away from table when ammonia is opened and to hold their noses. An adult should measure the ammonia and bluing (it stains). Mix all ingredients in jar until salt is dissolved.
4. Slowly pour mixture over coal pieces to saturate. Let children sprinkle drops of food coloring in separate areas on the coal.
5. Place dish where it can be observed but not disturbed.

GETTING READY:

Spread newspapers on work area.

Have smocks for children.

*Keep ammonia bottle tightly capped until needed. Open well away from the face. *Keep out of the reach of children.*

Crystals may begin to form on the coal within a few hours and continue to form for several days. The dissolved minerals move to the surface of the coal as the liquid evaporates and the mineral residue forms crystals. Liquid will continue to move to the surface of the newly formed crystals to form crystals on top of crystals. Some children may think they are growing as a plant grows. Point out that minerals can change, but only living things grow.

2. Will crystals form from salty water?

LEARNING OBJECTIVE: To observe the formation of crystals.

MATERIALS:

2 tablespoons rock salt (halite)

1 cup hot water

Glass pint jar

String

Pencil

SMALL GROUP ACTIVITY:

1. Pour hot water into the pint jar. Let children measure rock salt into water, stirring until dissolved.
2. Tie string around the pencil. Place it across the jar top so the string dangles in the salt solution.
3. Place jar on a sunny windowsill. Check the next day for beginning crystal formation. Look at the underside of the pencil. Watch for several days.

Read *Greg's Microscope,* by Millicent Selsam. If you have access to a simple microscope (40 × will do), follow Greg's method for forming salt crystals. Through the microscope tiny crystals can be seen as perfect, sparkling cubes.

PRIMARY INTEGRATING ACTIVITIES

Music (References in Appendix 1)

1. A familiar folk song can be easily adapted to help children remember that mountains are rock formations. Sing "The Bear Went Over the Mountain," substituting " . . . But all he could see was the rocky part of the mountain" for " . . . But all that he could see was the other part of the mountain." You could mention that some mountains are covered with soil and growing things while other mountains are bare rock on top.
2. Listen to "What's Inside Our Earth?" sung by Tom Glazer on the recording *Now We Know (Songs to Learn By).*

Math Experiences

Counting. The Grab Bag Game may be played with stones. Place a container of stones in the center of a circle of children. Let them take turns reaching in to scoop out as many stones as they can hold in two hands. Then count the results. Keep score with an abacus, a numbered spinner, or make dice from 4" (10 cm) squares of cardboard to determine how many rocks players win.

Buried Treasure. If a sandbox is available, bury a specific number of stones (or stones and shells, if you wish to include classifying) for children to hunt. A large sandbox may be "staked out" with string as separate territories for several children to use amicably.

Counting Cans. Mark numerals from 1 to 10 (or more) on the sides of low metal cans. Line up the cans in order on a table or shelf. Place a coffee can full of gravel beside them for independent counting of appropriate quantities of stones to put in numbered cans.

Ordering. Search through the bucket of gravel to find stones of conspicuously different sizes. Place them in a paper bag. Pass the bag among the children so they can choose a stone (with eyes closed). Allow time for much fingering and feeling in choosing. Ask for the smallest stone found. Next ask children to compare their stones with the smallest one to find a stone just a bit larger to put next to it. Build up to the largest stone. Older children who have had experience measuring with rulers could use them for this activity. Suggest ordering rocks according to texture, from smoothest to roughest, and color, from lightest to darkest.

Weighing. Keep two coffee cans full of stones near the classroom balance for independent use by the children.

Stories and Resources About Rocks and Minerals

BAYLOR, BYRD. *Everybody Needs A Rock.* New York: Charles Scribner's Sons, 1985. A young child's view of rock hunting.

______ . *If You Are a Hunter of Fossils.* New York: Charles Scribner's Sons, 1984. Lovely fragments of prose evoke images of the ancient plants and animals found as fossils.

BRANLEY, FRANKLYN. *Earthquakes.* New York: Harper & Row, 1990. Children will be impressed by the nature and power of earthquakes. They will be reassured by learning what they can do to take care of themselves, should they experience an earthquake.

______ . *Volcanoes.* New York: Harper & Row, 1985. Calming assurance that good warning systems keep people safe from volcano danger follows the information about this natural phenomenon.

COBB, VICKI. *Lots of Rot.* New York: J.B. Lippincott, 1981. Interesting experiments and light-hearted illustrations demonstrate the important role of decomposers in soil formation.

COLE, JOANNA. *The Magic School Bus Inside the Earth.* New York: Scholastic, 1988. The author and her illustrator, Bruce Degen, entertain delightfully as they present information about Earth's interior.

CURRAN, EILEEN. *Mountains and Volcanoes.* Mahwah, NJ: Troll, 1985. Lovely watercolors illustrate this very simple text. Amplify with information on how very slowly mountains change; how rarely volcanoes erupt. Paperback.

DEPAOLA, TOMIE. *The Quicksand Book.* New York: Holiday House, 1977. A lighthearted but factual explanation of the phenomenon to change this topic from menacing to interesting. Directions for making a pailful of quicksand are included. Paperback.

GANS, ROMA. *Rock Collecting.* New York: Harper & Row, 1984. Children are drawn into this book that says, "The oldest things you can collect are rocks . . . millions and millions of years old." Basic rock formation and cheerful illustrations are included. Paperback.

HOBAN, RUSSELL. *Baby Sister for Frances.* New York: Harper & Row, 1964. Frances keeps her collection of gravel in a can and uses it as a noisemaker.

______ . *Nothing To Do.* New York: Scholastic Book Services, 1969. Father gives Walter a stone, polished smooth by the river, calling it a "magic something-to-do stone."

KAUFMAN, JOE. *Joe Kaufman's Big Book About Earth and Space.* New York: Golden Books, 1987. This book compiles capsules of child-level information about the geology, geography, and weather of our planet Earth. A good classroom reference with basic answers to children's questions about earthquakes and volcanoes, along with facts about rocks, minerals, and crystals.

MARCUS, ELIZABETH. *Rocks and Minerals.* Mahwah, NJ: Troll, 1983. Part of this book takes us on a rock hunt, then helps us identify what we have found. A good reference for young collectors. Paperback.

MCNULTY, FAITH. *How to Dig a Hole to the Other Side of the Earth.* New York: Harper & Row, 1979. Earth science facts blend easily into this fantasy journey below the crust of the Earth.

PETERS, LISA. *The Sun, the Wind, and the Rain.* New York: Holt, 1988. This fine lesson in geology concepts parallels, page by page, the forming and changing of a mountain with a child's construction on a nearby beach of a sand mountain. Read this story aloud at the sand table or sandbox as you build and change a sand mountain along with the child.

PODENDORF, ILLA. *The New True Book of Rocks and Minerals.* Chicago: Children's Press, 1982. Simple statements about the three ways rocks are formed.

SANTREY, LAURENCE. *Earthquakes and Volcanoes.* Mahwah, NJ: Troll, 1985. Information for primary-grade children about these powerful natural events will reassure some, and raise concerns for others who live close to the Pacific plate. Paperback.*

SELSAM, MILLICENT. *Greg's Microscope.* New York: Harper & Row Junior Books, 1963.

WYLER, ROSE, & AMES, GERALD. *Secrets in Stones.* New York: Four Winds Press, Scholastic Magazines, 1970. This child-oriented book starts with what a child can see and find, then explains how rocks and rock formations came into being. A good reference book for younger children; a readable guidebook for older children.*

______ . *Science Fun with Mud and Dirt.* New York: Simon & Shuster, 1986. Among the simple activities Wyler describes as "mudpie science," are directions for making a miniature mud house and brickmaking.

Poems (References in Appendix 1)

Read the profound poem "Rocks," by Florence Parry Heide, in *Poems Children Will Sit Still For.*

*Starred references, written at the young child's level of understanding, can help teachers with minimal backgrounds in science expand their knowledge base.

Read the following poem with a box of your favorite rocks at hand:

My Rocks

Here in this box
I keep my rocks.
They came from under the ground.
Some are striped, some are plain,
Some are tiny as a grain.
My favorite is round!

Art Activities

Rock Sculpture. Let children create additive sculpture by joining rocks of assorted sizes, using white glue. The resulting sculpture forms may be painted with tempera. Sculptures that have been coated with shellac (adult work) can be used as paperweights. Older children enjoy trying to create fantasy by gluing dyed aquarium stones, seeds, and tiny shells to their sculptures.

Rock or Sand Collage. Small stones, painted if desired, can be glued in patterns onto heavy cardboard. Color sand by shaking it in a tightly covered jar with dry tempera paint. Punch nail holes in the lids to make sand shakers. Let the children paint swirls of diluted white glue on construction paper, using cotton swabs, then sprinkle colored sand over the glue patterns. Surplus sand can be tipped onto a small tray and reused.

Chalk Drawings. When children use chalk as an art medium, recall that the chalk material was once a rock.

Sand Painting. Prepare four colors of sand as you did for 2 above. Give children each a clear, screw-topped container. Let them spoon successive layers of sand into the container, being careful not to disturb the container. Probe gently along the inside of the container with a toothpick to produce interesting designs. My student, Peggy Robinson, uses these layers to help children understand how the time when dinosaurs lived is determined. Add one layer each day, recording the date and color of sand. Place a small chicken bone on one layer.

After 5 days ask, "Which is the oldest layer of sand? Which is the newest? How do you know?" Explain that the Earth's crust is made up of layers of rock that have built up on top of each other. Scientists can tell how old dinosaur remains are by the layer of rock in which they are found.

Clay. Remind children of its origin when using natural clay for art projects. Also, refer to Rose Wyler's book, *Science Fun with Mud and Dirt*, if making a model mud hut would extend a social studies concept.

Dramatic Play

Introduce rocks as a play material in the indoor or outdoor sandbox. Rocks can be formed in lines to make the floor plan of a house. They can also be piled up to make furniture, and sticks or clothespins can become the people who live in the house. Lines of rocks can also outline highways for toy cars to travel through sand-table dunes and deserts. In addition, children find relaxed enjoyment using sifters to separate the rocks from the sand at cleanup time.

Food Experiences

Children are surprised to find that we eat small amounts of one kind of rock every day (salt). Show them some rock salt if available. Mention that all the salt that is mined under the ground was once part of ancient oceans.

Another link between food and rocks is the use of rocks to grind seeds into meal and grain. Perhaps you can find a bag of stone-ground flour or cornmeal in a natural foods shop. (One teacher let fascinated children grind dried corn between two rocks as part of a study of American Indians.)

Field Trips

Walk around the school neighborhood to look for the following: rocks in their natural setting, rocks that have been cut for use in buildings, and products made from rocks and minerals. Look closely at the natural rock for signs of weathering, such as cracks or surfaces worn smooth by years of exposure to heat, cold, wind, and rain. Young children personalize familiar rocks according to the events they associate with them: our picnic rock, our sitting-on rock, our story-time rock, and our where-we-found-the-turtle rock. Use the special rock names when you can.

Look for old stone steps or stone curbings that have been worn down into sloping shapes by thousands of feet treading upon them. Look for monuments, stone walls, stone windowsills in old brick buildings, and flagstone or crushed limestone paths.

Man-made materials containing rocks and minerals are everywhere. Concrete or blacktop playground surfacings contain coarse or fine gravel bonded together with concrete or asphalt, both derived from rocks. Cement blocks, bricks, tiles, terrazzo flooring, porcelain sinks, glass, iron railings, steel slides, and fences around school playgrounds are made with rock or mineral raw materials. No school is without structures of this sort to visit. Talking about these strong, solid parts of their environment seems to contribute to children's feelings of security.

Make an ecological scavenger hunt the focus of a field activity, as Margaret Drysdale does. She takes her classes to an empty lot where they check the area for debris. They classify their collected materials into two piles: man-made materials and natural materials. She explains that only the things from nature can be left on the ground, because they will decompose and eventually become part of the soil. The class carries the man-made materials back to the school recycling/trash bin, because those things won't decompose. She emphasizes that a discarded aluminum can or plastic item could still be littering the land when the children are old enough to be grandparents.

Creative Movement

Take a rock walk. Children enjoy moving in a circle, interpreting movement suggestions and following the rhythm of the teacher's drumbeats or clapping. Adult participation in the activity encourages the involvement of children. Suggest ways to move in a story form such as this: "Let's go for a rock walk. Let's pretend to be barefooted. We'll start on this gravel path. Oh! It's *hard* to put our feet down on the broken rocks. We'll have to move quickly and lightly to the end of the path. Step, step, step, step. There, that's better. We've come to a sandy place. Let's scuffle our feet in the sand. Scuffle, scuffle, push up little piles of sand with each step. Oh look, we've come to a wet place where we have to walk on moss-covered rocks. They are so slippery to walk on; let's move very carefully with each step. Look at that huge rock. Let's try to climb it, bending way over to hold on with our hands. Climb slowly, slowly. Now we're at the top. Let's run down the other side to the beach ahead. Run, run, run to the sandy beach. But this sand is *hot* from the sunshine. We can't move slowly here. Let's move quickly down to the edge of the water to cool our feet. Now let's sit down to rest."

SECONDARY INTEGRATION

Keeping Concepts Alive

Since rocks and minerals are so commonplace, the topic would become tedious if reference were made to them at every opportunity. Do mention rocks when something a bit different prompts a comment, such as, "Is Lynn wearing a very special rock in her bracelet today? Tell us about it." Bring interesting rocks to share with the children whenever you come across one. Talk about its texture, color, or whatever appeals to you. The children will respond similarly with their favorites when they know of your appreciation of rocks.

Relating New Concepts to Existing Concepts

The inclusion of rocks as a material for experimentation is suggested for the topics of water, electricity, and the effects of gravity. It could be pointed out in discussing the effects of magnetism that the original source of magnetic material is a rock called lodestone (magnetite). Magnetite is found in this country near Magnet Cove, Arkansas.

The tie between soil formation and growing things is an easy relationship to mention. Wet sand is sometimes used as a growing medium for rooting stems such as begonias or pineapple tops. The powdered rock made when children pounded soft rocks could be used in a seed germinating experience. Compare the results with seeds growing in good topsoil containing humus and other rotted organic matter.

Discuss how thoroughly rotted natural materials improve the soil, so that healthy, strong plants will grow. Ask, "Do people use some plants for food?"

Talk about renewing the soil this way as one of the wonderful cycles of nature: from living plants to decomposition . . . from enriched soil to living plants and food again . . . and again . . . and again. Talk about saving and improving the soil with composting as one way that people help to renew our planet.

If possible, in early fall bury several non-biodegradable materials, such as a plastic spoon, a foam cup, or an aluminum can in a marked location on the school grounds. Also include a paper bag and a regular plastic bag. Dig them up in late spring to see if the biodegradable and non-biodegradable materials were affected differently in the soil. Encourage the children to make the connection between what they observed and our need to reuse and recycle man-made objects that do not decompose. Celebrate Earth Day, April 22.

Family Involvement

Invite families to share special rocks or fossils with the class. Masking tape name labels are helpful for insuring the safe return of borrowed rocks. Suggest that they point out to their children areas of rock exposed by highway construction or special rocks that are landmarks in the area. Win the hearts of parents by forewarning them to check the pockets of their children's jackets and jeans for rocks before laundering them, now that the children's interest in rocks has been whetted.

EXPLORING AT HOME*

Start a Soil Time-Capsule. Dig a shallow hole, about a foot square, in a little-used spot of ground near your home. Put a layer of leaves or grass clippings on one side of the hole; a piece of aluminum foil on the other side. Cover these materials with soil. Mark the space with rocks so you can find it and dig it up in one year. Circle the date on the calendar, so you won't forget to check the results. (You'll find that the natural materials have started to rot, and perhaps get moldy. The foil will not have changed.) Let your child tell you as much as possible about what happened to the two kinds of material, and why we need to recycle or reuse things that won't decompose in garbage landfill sites. Make this a family rule: "We don't waste and we don't litter. We save the Earth and make it better."

Help a Rockhound. If the child in your home is an avid rock collector, help him or her enjoy organizing and thinking about these wonderful finds. Ask your child to tell you about favorite rocks: what is special about their appearance, and where they came from. Help your child identify the favorite rocks, using a library book, or buying an inexpensive paperback nature guidebook like *Rocks*

*These suggestions may be duplicated and sent home to families.

and Minerals, by Zim and Shaffer. Try to find a place to display as many rocks as possible. An egg carton makes a simple display box. Use the surplus rocks around the house in different ways: in the bottom of a soap dish to keep the soap dry; under potted plants; outdoors under a downspout.

RESOURCES

ASIMOV, ISAAC. *How Did We Find Out About Earthquakes?* New York: Walker, 1978.

COMMONER, BARRY. *Making Peace with the Planet*. New York: Pantheon Books, 1990.

EARTHWORKS GROUP. *50 Simple Things Kids Can Do to Save the Earth*. Kansas City: Andrews & McMeel, 1990. Paperback.

GATES, RICHARD. *Conservation*. Chicago: Children's Press, 1982. Includes the makeup of soil and damage people have done to the environment.

HYLER, N., SUTTON, F., KEEN, M., & ROBBIN, I. *Rocks and Minerals*. New York: Grossett & Dunlap, 1982. Rock formation and identification.

MARCUS, ELIZABETH. *Rocks and Minerals*. Mahwah, NJ: Troll, 1983. Paperback.

SABIN, LOUIS. *Fossils*. Mahwah, NJ: Troll, 1984. General information about fossils and paleontology.

WILLIAMS, R. A., ROCKWELL, R. E., & SHERWOOD, E. A. *Mudpies to Magnets*. Mt. Rainier, MD: Gryphon House, 1987. Soil experiments, brickmaking.

ZIM, HERBERT, & SHAFFER, PAUL. *Rocks and Minerals*. New York: Golden Books, 1989. Comprehensive pocket guide for rock identification. Paperback.

Teaching Resources

NATIONAL SCIENCE TEACHERS ASSOCIATION. *Earthquake Curriculum*. Federal Emergency Management Agency, 1988. This extensive K–6 multidisciplinary curriculum actively involves children in learning about the nature of earthquakes, recognizing an earthquake, and earthquake safety and survival. Free of charge by writing to: Susan McCabe, FEMA SL-NT-EN, 500 C Street SW, Washington, DC 20472.

CHILDREN'S TELEVISION WORKSHOP. The *BIG BIRD Get Ready for Earthquakes* kit. Earthquake safety information including a board game and an audiocassette of a song, "Beatin' the Quake." $2.25 for a single copy. Check payable to: Children's Television Workshop, Dept. NH, One Lincoln Plaza, New York, NY 10023.

FEDERAL EMERGENCY MANAGEMENT AGENCY. *Coping with Children's Reactions To Earthquakes and Other Disasters*. FEMA #48. Single copies free. Write to: FEMA, P. O. Box 70274, Washington, DC 20024.

11

The Effects of Magnetism

Magnets attract iron, steel, *and the attention of young children.* While magnetic attraction cannot be seen or felt, *its effects can be seen and felt.* Children can accept the reality of an invisible force when they can have experience putting that force to work. The following concepts underlie the experiences in this chapter:

- Magnets pull some things, but not others
- Magnets pull through some materials
- One magnet can be used to make another magnet
- Magnets are strongest at each end
- Each end of a magnet acts differently

In this chapter children will experiment with familiar objects to see what magnets will pull and what they will pull through, make temporary magnets, and discover how opposite magnetic poles affect each other.

CONCEPT: Magnets pull some things, but not others

1. What will magnets pull?

LEARNING OBJECTIVE: To experience visually and tactually the effects of magnetic attraction and to apply the effects to sort iron or steel objects from nonferrous objects.

MATERIALS:

Magnets of assorted shapes and sizes

Small foam meat tray filled with test items for each child. (Iron/steel suggestions: keys, key chains, bolts, screws, nails, paper clips, lipstick cases);

SMALL GROUP ACTIVITY:

1. Give each child a paper clip and a small magnet to try out. To see the effect well, slide the magnet toward the clip on the table top. To *feel* the effect, hold clip in the palm of the hand.
2. "Do you think the magnet will pull everything to it? Let's find out with the things on the small trays."
3. After some exploration, suggest sorting objects pulled by the magnets from those that are not. Use the *yes* and *no* trays.

(Noniron/steel items: pennies, brass fasteners, rubber bands, plastic, glass, wood, aluminum objects)

Two large trays or box covers

GETTING READY:

Have only magnets on the table when children gather. Count them together. (Tiny magnets are easily misplaced.) Prepare a mixed collection of objects for each pair of children.

Tape a *yes* label on one tray and a *no* label on the other.

4. "Things on the *yes* tray do not look alike, but they are made of the same stuff." Children may note that all items are metal. You can use the specific terms *iron* or *steel.* "Magnets pull only on the iron or steel objects." (Magnets also attract cobalt and nickel, two minerals not commonly used alone in manufactured articles.)

Note: Toy magnets are rarely strong enough for school use. School suppliers and scientific equipment suppliers sell sturdy magnets. An electric motor repair shop may be willing to give you an old magnet removed from a motor. (Remove your watch before handling a very strong magnet of this size.) Scientific equipment outlet stores carry as many as 40 shapes (rings, cylinders, bars, horseshoes) and kinds (steel, ceramic, rubber) of magnets. See appendix 4 for catalogs.

Read *Mickey's Magnet,* by Franklyn Branley and Eleanor Vaughn.

CONCEPT: Magnets pull through some materials

1. Can magnets pull through things that are not attracted?

LEARNING OBJECTIVE: To discover forms of nonferrous materials through which magnetism will pass.

MATERIALS:

Magnets*

Steel wool pad

Iron and steel objects (nails, washers, bolts, clips)

Paper, cardboard, aluminum

Shoe box

Drinking glasses

Sand or dirt

Water

SMALL GROUP ACTIVITY:

1. "Do you know what this (steel wool) is made of? Could a magnet help you find the answer?"
2. Tear off bits of steel wool. Give some to each child. Find out if a magnet can pick up steel that is covered with a piece of paper. Try cardboard and aluminum. "Will magnetism pull through these things to pick up steel?"
3. Touch the magnet to the outside of a dry tumbler. Does it attract the steel object inside the tumbler? Try different magnets and different objects.
4. Suggest dipping magnet into the sand-filled box and the tumbler of water. Can magnetism pass through materials which it does not attract?

Michelle delights in the iron filing patterns she creates as her big magnet attracts them through cardboard.

GETTING READY:

Put steel object in tumbler.

Put steel objects in box; cover with sand.

Put water in another tumbler; drop a washer in it.

Put water in a third tumbler; drop a washer in it.

*The thickness of the noniron material and the strength of the magnet are factors in the success of this experiment. A strong, new magnet will attract a paper clip through one's fingertip. A weak magnet will not attract a paper clip through cardboard. To keep magnets strong, always store two bar magnets together with opposite poles touching (north to south and south to north) when not in use.

Group Discussion: When the children try out various magnets they will find that some work better than others. Talk about this with the whole group. Children can help keep magnets strong by remembering to put a steel "keeper bar" across both ends of a horseshoe magnet. They can also keep magnets strong by not jarring them.

Note: A child may believe that a keeper bar is also a magnet—one that just doesn't happen to work right when it is separated from the magnet. From the child's experience a plain bar of metal has no assigned function. Find a familiar iron or steel object to use as a temporary keeper bar. "Children's scissors are not magnets. Will a magnet attract them? If so, then we can leave this pair of scissors across the ends of the horseshoe magnet to keep it strong, while you experiment with the other keeper bar." Experience may clarify a child's ideas about the keeper bar, while verbal persuasion from an adult may not.

CONCEPT: One magnet can be used to make another magnet

1. Can we make a magnet?

LEARNING OBJECTIVE: To learn how to make a temporary magnet.

MATERIALS:

2" (5 cm) needles or straightened paper clips

Strong magnet (bar, horseshoe)

Paper clips, steel wool, steel straight pins (not brass pins)

SMALL GROUP ACTIVITY:

1. Let children use the magnet to determine if needles or clips are made of steel.
2. "Try to pick up steel wool bits with the needles or clips. Are they magnetic?" (They won't be yet.)
3. Show children how to pull the needle *across one end* of the magnet *in one direction.* Count aloud about 25 strokes.
4. "Now try to attract steel wool with the needle. It should be magnetized. It has become a *temporary magnet.*"

Note: This is a good time to demonstrate how jarring a magnet weakens it. Hit the magnetized needle against something hard a few times. Now try to pick up some light steel object. The pull will be very weak.

CONCEPT: Magnets are strongest at each end

1. Which parts of a magnet are strongest?

LEARNING OBJECTIVE: To notice that the ends of a magnet act differently from the middle of the magnet.

MATERIALS:

Bits of steel wool

Paper clips, steel key chain, or light switch pull chain

Horseshoe and bar magnets

GETTING READY:

Cut the steel wool into fine bits. (Did some of the steel cling to the scissor edges? The scissors became a temporary magnet.)

Open the key chain full length.

SMALL GROUP ACTIVITY:

1. Hold the curved end of the horseshoe magnet over steel wool. Notice if any steel is attracted to the middle of this magnet. "Hold the *ends* of the magnet over the steel wool. Do the *ends* act differently from the middle?" Try with a bar magnet.
2. Touch the chain with both ends of a magnet. Notice whether the whole chain clings to the magnet or if the middle part dangles free. (Chain should be longer than the bar magnet, 3" (7.5 cm) key chain.)
3. Dangle the key chain $\frac{1}{4}$" (1 cm) above the center of a bar magnet. (The pull of a strong magnet will visibly curve the end of the chain toward one end.)

CONCEPT: Each end of a magnet acts differently

1. Are magnets alike on both ends?

LEARNING OBJECTIVE: To notice that each end of a magnet acts differently, and that like ends of two magnets push away while unlike ends pull together.

MATERIALS:

2 strong bar magnets or 2 horseshoe magnets

String

Nail polish or tape

GETTING READY:

If using lightweight magnets, tape the string to a tabletop or chair seat.

Tie the string to any horizontal bar or chair back that will allow a heavy magnet to swing freely.

SMALL GROUP ACTIVITY:

1. Tie the string to the center of a magnet, balancing it.
2. Let the magnet hang and swing freely. When it stops moving, one end of the bar (or one side of the horseshoe magnet) will be pointing north. Mark this end with tape or a dot of nail polish. Do the same for the second magnet. (This is also how a magnetic compass works. Some children may know about compasses already. Others may find directional discussion confusing at this point.)
3. Let the children hold a magnet in each hand, then try to touch like ends (north to north, south to south). "What do you feel? Try to touch unlike ends. What happens?"

Give children lots of time to explore the attraction and repulsion effect. With strong magnets, results are fascinating. This understanding will clarify the jumping effects of "Magniks" sold by Creative Playthings. To retain strength, make a point of storing bar magnets in pairs with opposite poles touching.

Note: In this instance, using the correct scientific term can create more confusion than clarity. The ends of magnets are called the north (seeking) pole and south (seeking) pole. Adults find it confusing that the North *(geographic)* Pole is not in quite the same location as the Earth's North *(magnetic)* Pole. Young children hold firmly to yet a different abstraction about the North Pole.

PRIMARY INTEGRATING ACTIVITIES

Music

Mention to children that we couldn't hear sounds from a stereo, VCR, radio, television set, or tape recorder without magnets. Recording tape is coated with magnetized powdered iron, and the other four devices have magnets in their mechanisms.

Try singing the "Magnet Song" (page 200).

Math Experiences

Magnet in the Grab Bag. Half fill a large grocery bag with used bottle caps, or use 3 or 4 pounds of common nails instead. Tie a strong magnet to a stick with

string, fishpole style. Let children take turns dipping the magnet into the bag, then counting aloud the number of caps they have pulled up. A numeral recognition version of this game can be played by cutting out colored paper fish, writing a numeral on each, and fastening a paper clip or safety pin to each fish.

Plastic numerals and counting shapes are made with magnets to use on coated steel bulletin boards. Use them to form sets of objects, labeled with corresponding numerals. Make a game of this by letting children draw the numerals from a paper bag, choose the corresponding quantity of objects, and place them on the bulletin board. You may have an old tin-coated steel baking sheet in your kitchen that would substitute well for the commercial steel bulletin board.

Stories and Resources for Children

ADLER, DAVID. *Amazing Magnets.* Mahwah, NJ: Troll, 1983.

BRANLEY, FRANKLYN, & VAUGHN, ELEANOR. *Mickey's Magnet.* New York: Thomas Y. Crowell, 1956.

CHALLAND, HELEN. *Experiments With Magnets.* Chicago: Childrens Press, 1986. Carefully presented information and verifying experiments, including one that demonstrates the opposite poles in round magnets.*

PODENDORF, ILLA. *Energy.* Chicago: Childrens Press, 1982. Includes a brief description of magnetic energy.

SANTREY, LAURENCE. *Magnets.* Mahwah, NJ: Troll, 1985.

Storytelling with Magnets

Tell a story featuring practical uses of magnetic gadgets that could also be props for the telling. A small magnet could save the day when father drops his keys down the register, when sister's box of bobby pins spills in her bubble bath, when someone upsets a box of pins, a glass jar of nails breaks on the garage floor. These and other calamities could be resolved with an upholsterer's tack

*Starred references, written at the young child's level of understanding, can help teachers with minimal backgrounds in science expand their knowledge base.

FIGURE 11–1

hammer, a magnetic-holder flashlight, a paper clip or pin box with a magnetized top, or other magnetic equipment that may be available.

The effects of magnetism can be used throughout the school year as a storytelling device in the following three ways:

Magnetically Directed Puppets. Hammer several tacks or a pronged steel caster to the bottoms of small wooden dollhouse dolls. Put an open shoe box on its side to make a platform for the puppets. (Figure 11–1.) Make the puppets move with magnets held in each hand beneath the platform. (Puppets can be thread spools with button heads glued on, clothespin dolls with paperclip feet, or pipe cleaner dolls.) Children love using magnet puppets to retell familiar stories or to create their own plays.

Paper Doll Stories. Cut out paper figures to illustrate a story. Tape paper clips or small safety pins to the backs. Invert a large grocery bag (plain or decorated) as a backdrop, and manipulate the paper figures with magnets held inside the bag.

Magnetic Bulletin Board Stories. Use fabric or paper figures as you would for a flannel board story. Hold them in place with a shirt button-sized magnet. Try to fit one of the whimsical magnet insects into the story.

Art Activities

Make junk sculpture with iron or steel.

1. Prepare bottle caps and cocoa tin lids by punching two nail holes into each one.
2. Prepare sculpture bases using mounds of damp clay, chunks of foam, or cardboard fruit trays.
3. Put out a tray full of iron and steel discards (paper clips, nails, bottle caps, hairpins, pipe cleaners, cocoa tin lids, twist bag closures, washers, strips of screening, soft wire). Put out several magnets so that the children can test

materials for magnetic attraction. Explain that the sculpture for today will be made only of iron or steel.

4. Show how wire can be threaded through the punched bottle caps, how paper clips can be opened, and how wire pipe cleaners can be coiled around a finger to make interesting shapes.

Directions for magnetically manipulated painting with steel objects appear in *Mudpies to Magnets,* (see Resources).

Dramatic Play

Automobile Service. Tie a small magnet to a toy tow truck for children to use in hauling steel cars to the garage.

Table Games Using Magnets

Name Game. Print capital letters on separate small squares of paper. Put a staple through each square. Pile the squares into a small basket. Let the children fish for those letters contained in their names with a magnet tied to a stick on a string.

Magnet Construction Game Box. Collect three or four small, strong magnets with keeper bars (or nails to serve that purpose). Find iron and steel discard items similar to those suggested for the junk sculpture. Try to include steel key chains, notebook rings, old keys, and cocoa box lids. Children enjoy combining odd shapes that are held together by magnetic attraction. Try to find a tin cookie box for storing the game. The lid and the box can serve as bases for the constructions.

Commercial magnetic building sets are available through school equipment catalogs. Magnetic games, magnet sculpture sets, and "Magnet Marbles" are available in toy and speciality stores. The Nature Company offers such items in their mail-order catalog (see Appendix 4).

Thinking Games: Imagining

I Am a Magnet. Collect a tray of assorted objects: key, bottle cap, spool, stick, rock, nail, bolt, paper clip. Sit with the children in a circle on the floor. Place the tray in the center. Start off as the leader when introducing the game. Ask the child next to you to choose one of the objects to pretend to be. Say, "I am a great, strong magnet. What are you, Maria?" If Maria replies that she is a bottle cap, say, "Then we'll cling together," and hold her hand. Encourage Maria to say to the child next to her, "Josh, I'm a magnetized bottle cap. What are you?" If Josh decides to be a rock, Maria goes on to the next child until she finds someone to cling to. End the game with, "Now I am a teacher again and my pulling strength is gone."

SECONDARY INTEGRATION

Keeping Concepts Alive

Point out magnets in use around the school whenever they come to your attention: refrigerator door sealing strips, car seat belt clasps, fancy buckles on children's belts, cupboard door latches. For the latter example, it is a good idea to suggest that children using those cupboards try to close them gently, without banging them, so that the magnets won't lose their strength.

Make magnets a standard part of classroom cleanup equipment. Hang a magnet and keeper bar on a low peg near the workbench if you have one. Let the children use it on the floor, in workbench drawers, or on the workbench to gather up stray nails. Use it to sort out tiny nails from wooden shapes after children have worked with hammer-on design kits. Use the magnet to locate small steel cars that have become buried in the sandbox.

Relating New Concepts to Existing Concepts

1. When discussing the effect of magnetism passing through materials, ask the children whether air is one of the substances that magnetism passes through. How can they tell? Recall with the children that even though we cannot see air or magnetism, we have discovered that both are real things.
2. Set up a water play game that depends upon magnetism's passing through water. Let the children make barge-shaped boats from pressed foam, and fasten a paper clip to it with a rubber band. Stack three blocks under each end of a shallow cake pan. Fill the pan with water and launch the foam boats on it. Children can guide their boats through the water with magnets held beneath the pan.
3. Combine buoyancy and the principle of attraction and repulsion of like magnetic poles in a water play activity. Help the children magnetize 2" (5 cm) blunt needles, then place them on top of barge-shaped foam boats. Float the boats in a pan of water. Push the boats ahead by approaching the end of the needle with the like pole of a magnet; pull boats back with the unlike pole of the magnet. If a needle rolls off a boat, the children can go fishing for it in the water with a magnet.

EXPLORING AT HOME*

A Scrap-Dance Box. Have fun putting together a magnet toy with your child. You'll need a small, but strong magnet of any type, a shallow box, some plastic wrap, tape, and a steel wool pad. Shred very small bits of steel wool from the scrubbing pad. Heavy kitchen shears work best for this job. Cut enough steel

*These suggestions may be duplicated and sent home to families.

wool scraps to barely cover the bottom of the box. Stretch the plastic wrap over the top to make a cover, and secure it to the box with tape. Together, enjoy making the scraps dance and creating patterns by pulling the magnet beneath the box.

A Hidden Magnet Hunt. Go on a hidden magnet hunt through your house with your child. Discover useful, but unseen magnets in paper clip holders, cupboard door catches, flashlight holders, message holders. Examine the magnetized plastic strip that holds the refrigerator door tightly shut. Talk about the invisible magnets that are important parts of car motors, electric motors, radios, telephones, television sets, tape recorders. Show your child the magnetized strip that you slide into automatic bank teller machines to activate them, if you have access to such equipment. Let your child know that we *need* magnets and use them every day!

RESOURCES

BERGER, MELVIN. *The Science of Music.* New York: Thomas Y. Crowell, 1988. Learn how microscopic magnets turn into audiotapes, pp. 123–124.

LEVENSON, ELAINE. *Teaching Children About Science.* Englewood Cliffs, NJ: Prentice-Hall, 1985.

WILLIAMS, ROBERT, ROCKWELL, ROBERT, & SHERWOOD, ELIZABETH. *Mudpies To Magnets.* Mt. Rainier, MD: Gryphon House, 1985. "Magical Magnet Masterpieces" art project, p. 87.

12

The Effects of Gravity

Children who have felt the tug of invisible magnetic attraction with their own hands can move from this awareness to simple understanding about the effects of gravity. These children are ready to put credence in the far stronger invisible pull that holds people, houses, and schools on the ground—the force of gravity. The learning experiences in this chapter amplify one central concept:

- Gravity pulls on everything

Drawing children's attention to an ever-present effect that is rarely noticed or labeled calls for more preliminary description than we usually offer to action-loving children. The first suggested gravity experience is a story that provides basic information. The active experiences involve measuring and comparing gravity's pull on objects, trying a pendulum, and bringing things into balance.

Lead into the gravity story by crushing a piece of paper into a ball and holding it between your fingers. "What will happen to this ball of paper if I open my fingers?" Find out if the children's predictions are correct. "Do you think the ball might do the same thing another time, or might it fall up instead? Did you ever stumble and fall up? Here is a story about the reason for this."

Use a flannel board and felt figures to illustrate your story. It could be stated something like the following:

> Once a girl had a dream. Everything seemed very strange. The girl was floating in the air looking for her house. She saw her friend floating close by with a ball in his arms. They wanted to play catch, but when the boy threw the ball, it drifted up out of reach. Then the girl saw her house bobbing up and down gently in the breeze. Her mom was very upset because somebody had spilled milk all over the ceiling. When the girl woke up, she was glad that her house was standing still.
>
> Something important was missing in that dream. Houses and balls and children don't float around. Milk doesn't spill up! There is a reason why those things don't happen. There is a very powerful force pulling down on everything in the world: the force called Earth's *gravity*. We can't touch gravity or see it. We can just see what gravity does.

Can you think of something else invisible that pulls some kinds of things? It reminds us of the way magnets pull on iron and steel. But gravity is much, much stronger than magnetism. Gravity pulls from inside the Earth. *It pulls on everything all of the time.* We are so used to it that we don't even notice it happening.

(Show the picture of an astronaut in his space suit.) The astronauts who walked on the moon had a strange experience. They had to wear very heavy boots to stand on the moon, because the moon's gravity pull is weaker than the Earth's gravity pull.

You can find out how much Earth's gravity is pulling on you at the science table.

CONCEPT: Gravity pulls on everything

1. Can we experience gravity pulling on us?

Invite a volunteer child to be the demonstration subject to introduce this activity. Have the child sit on a straight chair with both feet flat on the floor, the back touching the chair back, and the hands in the lap. Now ask the child to try to stand up without swaying her or his body forward, nor moving any other muscles. . . . not even a tiny bit. Encourage the child to try very hard to stand up, then report to the class what she or he is experiencing. Ask, "What will you have to do in order to stand up? Think about it."

Let the rest of the class take turns trying the experience in their small groups, while others observe for motionlessness. Afterward discuss what the children noticed. "What do you think held you to the chair when you didn't move a muscle?" "What did you need to do to stand up?"

Guide the discussion to the conclusion that it is gravity's force that holds us down in the chair. The force of gravity always pulls on everything. We have to use our energy to push ourselves out of the chair and stand up. It takes energy to move the muscles we use to stand up. It always takes energy to work against gravity's pull.

2. How much does gravity pull on different objects?

LEARNING OBJECTIVE: To experience the way gravity pulls on one's body.

MATERIALS:

Bathroom scale

Kitchen scale

4 small, sturdy bags

Sand, pebbles

Market bag with handles

Containers, scoops

Long mirror, if possible

GETTING READY:

Fill each small bag with 2 pounds of sand or stones.

SMALL GROUP ACTIVITY:

1. "When we weigh ourselves on a scale we are finding out how much gravity pulls on us."
2. Find out how much gravity pulls on each child. Record the weights.
3. Weigh each child again while he holds a sand bag in each hand. Compare with first weight record. Which way does gravity pull more?
4. Put both sand bags in the market bag. Have children hold it in one hand so the weight is on one side. Weigh each child again.
5. "Look in the mirror. Are you standing up straight now or leaning over? You lean away from the heavy side to keep your balance."

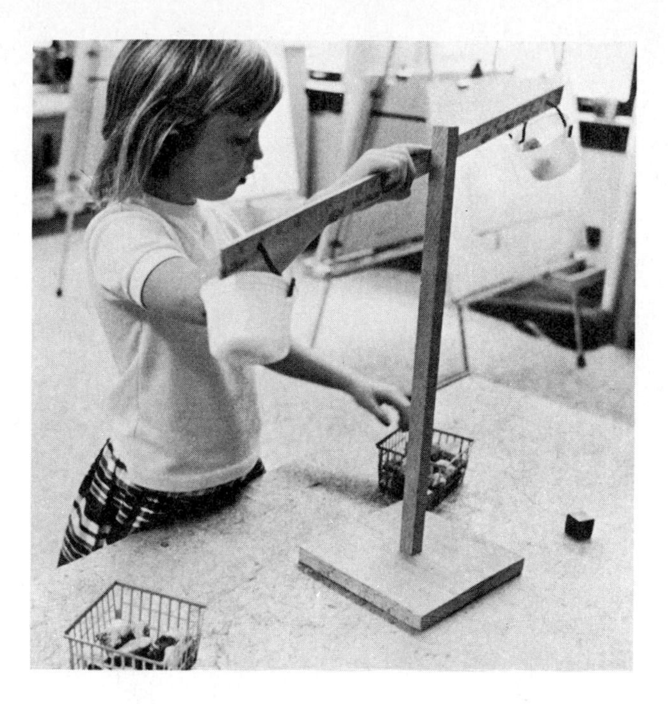

Chris uses the yardstick balance she helped assemble from scrap lumber.

Let children weigh objects on both scales. Containers of sand make good materials to weigh. Provide objects of varying weight and size. Let children discover that size does not always determine weight.

Group Discussion: Recall the earlier discussion about gravity holding everything down. Encourage children to share ideas about things that they see being pulled down from the sky by gravity (leaves, seeds, snowflakes, raindrops).

Mention that some things go up and move through the air for a long time (planes, birds, gliders, and so on). Wind, motors, and jet engines can lift planes up and keep them up for a while. Can the children lift themselves all the way off the ground for a little while by using their muscles to push against gravity?

Read "How Birds Fly" from *Great Big Air Book*, by Richard Scarry.

3. *Can we compare gravity's pull on objects?*

LEARNING OBJECTIVE: To explore ways of achieving equilibrium using varied small objects as weights in balance cups.

MATERIALS:

Scrap lumber: 2′ (60 cm) piece of dowel or broomstick; 8″ (20 cm) × 8″ (20 cm) piece of shelving

Nails: 6d and 4d sizes

Yardstick (meterstick)

Pipe cleaners

Small paper cups

SMALL GROUP ACTIVITY:

1. To make base: Mark center of shelving square. Nail dowel perpendicular to center of base. (Let children do the hammering.)
2. Nail yardstick (meterstick) loosely near the top of the upright stick, driving a 4d nail through hole at the 18″ (50 cm) mark. Yardstick (meterstick) should swing easily.

Shannon and Danielle compare the weights of shells and checkers on the old drugstore balance scale.

Small objects to weigh: washers, bottle caps, etc.

GETTING READY:

Drill holes or saw small notches 3" apart on yardstick (8 cm on meterstick) markings.

Make a hole at 18" (50 cm) and mark with the 6d nail.

Make 2 holes on opposite sides of each paper cup near top.

3. Insert pipe cleaners into yardstick (meterstick) holes. Hook them into paper cups. (Put one cup on each side. Let children add more cups as they wish. If notches were made instead of drilled holes, make pipe cleaner handles for cups and hang in notches.
4. Let children fill cups with small items to balance.

Provide materials as different in weight as cotton balls and bottle caps. Do five cotton balls balance five bottle caps? Is the balance arm lower on the cap side? What does this mean? (Gravity pulls more on caps, thus we say they weigh more than the cotton balls.) A balance can be made from a paper towel tube by following directions given in *Adventures With A Paper Tube,* by Harry Milgrom.

Try to keep the balances and containers of weighing materials accessible to the children for as long as they show interest in them. Put them away for a while; bring out periodically.

Group Discussion: Tie a small doll at the end of a foot of string. Hold the other end of the string, letting the doll dangle. "How could I make the doll go higher without lifting up the string? Good, I could swing the doll higher. Look, it is going high now, but it doesn't stay high. Why? Let's watch what happens when I stop swinging the string. Is it moving up as high now? It's going slower and lower, and it has almost stopped swinging. Is something pulling it?"

"Is this what happens to you when you're on the swing? What happens to you when you start to pull and push with your muscles? What happens when you stop working your muscles that way? Are you close to the top of the swing set or close to the ground when the swing stops? Why?" Tape the doll and string pendulum to the edge of a table for children to try later.

Read *Gravity At Work and Play,* by Sune Engelbrektson.

4. Can we keep objects in balance?

LEARNING OBJECTIVE: To explore ways to achieve equilibrium by placing rocks on a ruler which is balanced on a block.

MATERIALS:

12" ruler

Half-circle blocks

Small rocks

Tinkertoys

Desirable: Acrobat balance toy from India, sold by mail order gift shops

SMALL GROUP ACTIVITY:

1. Gently place a ruler on a block, with the 6" mark resting on top of the curve. "Let's see if the ruler will tip or balance. Can you try this?"
2. "Can we balance two rocks on the ruler?" Show how to slide a heavier rock toward the center (resting point) to balance a lighter rock at the other end of the ruler.
3. Make a platform of 2 Tinkertoy wheels and one long stick. Insert 2 sticks in a connector wheel and add small connectors to the stick ends. Rest connector wheel on the platform edge as shown in Figure 12–1.
4. Remove one stick, then the other, from the connector wheel. What happens? Is the connector wheel stable without them? (The location and weight of the sticks change the way gravity pulls on the connector wheel.)

REMEMBER

Respect the questions children ask.
Let your question encourage finding out.

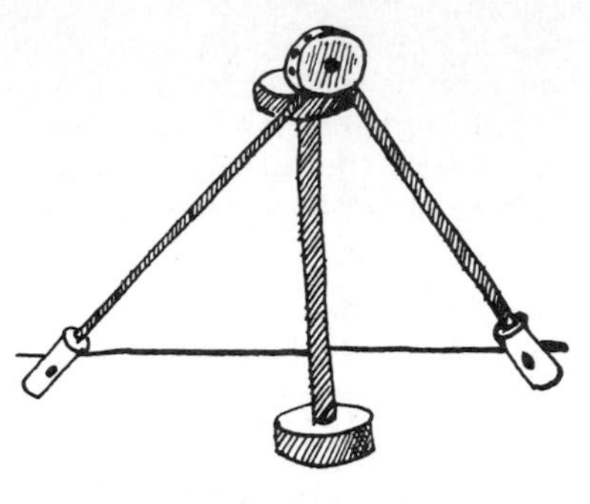

FIGURE 12–1

PRIMARY INTEGRATING ACTIVITIES

Music

When children are investigating the effects of gravity, they enjoy singing three nursery rhymes as though gravity were not operating. "Jack and Jill Went Up the Hill" might be sung ". . . Jack fell *up*, and broke his cup, And Jill bent over with laughter." "London Bridge" might be sung ". . . London Bridge is *floating up*," and "Ring Around the Rosy" might be sung ". . . One, two, three and we fall *up* in the tree."

Here is a gravity song to sing.

Math Experiences

When they experiment with a pan balance and weights, children can strengthen their understanding of numerical equivalence. Sets of standard weights are available commercially. In some types, the size and weight are related. To devise substitute weights fill screw-top plastic or metal containers with sand and gravel to 1-, 2-, and 4-ounce (25-, 50-, and 100-gram) weights as desired. Mark the corresponding numeral on each container.

Number balance equipment is made in several forms. One type uses weighted plastic numerals to make the mathematical equation balance as the weight balances. Keep a set of weights along with boxes of materials to weigh next to a pan balance for children to use when time permits.

Stories and Resources for Children

BARTON, BYRON. *I Want To Be an Astronaut.* New York: Thomas Y. Crowell, 1988. The sparse text and child-like illustrations leave room for group discussion about gravity, and may inspire children's creative expression at the easel.

BOONE, EMILIE. *Belinda's Balloon.* New York: Alfred A. Knopf, 1985. Belinda's balloon lifts her off the ground. Her sister helps solve the problem of becoming heavy enough to drift back to earth.

BRANDT, KEITH. *Air.* Mahwah, NJ: Troll, 1985. The section on air pressure illustrates the effect of Earth's gravity on the weather. Paperback.

BRANLEY, FRANKLYN. *Weight and Weightlessness.* New York: Harper & Row, 1971. Starting with gravity's effect on the child, this book ends with astronauts experiencing zero gravity in outer space.

______. *Gravity Is a Mystery.* New York: Harper & Row, 1986. This book helps children learn what gravity does. It doesn't explain just what gravity is, because no one has been able to explain it yet. Paperback.*

______. *Rockets and Satellites.* New York: Harper & Row, 1987. A simple experiment explains a rocket's lift against gravity. Use of satellites is illustrated. Paperback.

BRIGHT, ROBERT. *Georgie and the Magician.* New York: Doubleday, 1966. Georgie is a gentle ghost who does some floating in a magician's show. One involved child in my class said, on hearing this, "Now I know ghosts aren't real, 'cause gravity pulls on everything real."

COBB, VICKI. *Why Doesn't the Earth Fall Up?* New York: E. P. Dutton, 1988. Gravity and the laws of motion are explained simply.

CONAWAY, JUDITH. *More Science Secrets.* Mahwah, NJ: Troll, 1987. Includes directions for making a "whirling gravity ball" to illustrate how the sun's gravity holds Earth in its regular orbit. Paperback.

ENGDAHL, SYLVIA. *Our World Is Earth.* New York: Atheneum, 1979. This imaginary exploration of other planets shows astronauts floating in space away from the gravitational pull of planets.

ENGELBREKTSON, SUNE. *Gravity at Work and Play.* New York: Holt, Rinehart & Winston, 1963. This is a good first book about gravity for young children.

KAUFMAN, JOE. *Joe Kaufman's Big Book About Earth and Space.* New York: Golden Books, 1987. Two colorful pages on gravity are among the capsules of child-level information about earth science in this reference.

MCNULTY, FAITH. *How to Dig a Hole to the Other Side of the World.* New York: Harper & Row, 1979. Weightlessness is experienced at the center of the Earth in this fantasy journey.

MINARIK, ELSE. *Little Bear.* New York: Harper & Row, 1961. Chapter 3, "Little Bear Goes to the Moon."

MOCHE, DINA. *If You Were An Astronaut.* New York: Golden Books, 1985. Daily life aboard a space shuttle is portrayed. Teachers need to specify that "Zero G" means no gravity pull. Paperback.

MONCURE, JANE. *Skip Aboard a Space Ship.* Chicago: Children's Press, 1978. Two children take a fantasy space flight to the moon. A nice stimulus for dramatic play.

REY, H. A. *Curious George Gets a Medal.* New York: Houghton Mifflin, 1957. George takes a rocket ride into outer space.

RIDE, SALLY, & OKIE, SUSAN. *To Space and Back.* New York: Lothrop, 1986. Young children will enjoy hearing the portions of this book portraying the peculiarities of doing ordinary life routines weightlessly in an orbiting space capsule. Excellent color photographs of the astronaut/author in space.

SCARRY, RICHARD. *Great Big Air Book.* New York: Random House, 1971. "A Trip to the Moon" shows moonwalk problems.

*Starred references, written at the young child's level of understanding, can help teachers with minimal backgrounds in science expand their knowledge base.

WEBSTER, VERA. *Science Experiments.* Chicago: Children's Press, 1984. Easy experiments demonstrating gravity principles are included.

Swinging, sliding, blockbuilding, and falling down are part of many other children's stories as well. Casually refer to the effects of gravity's pull whenever possible.

Poems (see Appendix 1)

Gravity doesn't seem to operate in "The Folks Who Live in Backward Town," by Mary Ann Hoberman (in *Poems Children Will Sit Still For,* de Regniers, Moore, and White). Also read the classic poem about swinging, "The Swing," by Robert Louis Stevenson in *A Child's Garden of Verses.*

Art Activities

Use drinking straws, thread, and pressed foam meat trays to make simple mobiles. Children will need some help with this project, yet they can use their own ideas in a satisfying way.

Advance preparation for each mobile. Sew a double-knotted thread up through the exact center of the straw length. Tie thread into a 6" (15 cm) hanging loop. Sew an 8" (20cm) single thread down through the straw, $\frac{1}{2}$" (1 cm) from one end; sew a 6" (15 cm) single-knotted thread down through the straw $1\frac{1}{2}$" (4 cm) from the other end.

Give each child a 4" (10 cm) square and a 2" (5 cm) square of pressed foam to shape with scissors. Offer crayons for decorating the shapes. Older children can then punch a hole through the top of each shape, tie the 6" (15 cm) thread through the large shape, and the 8" (20 cm) thread through the small shape. Younger children can join shapes and threads with pieces of tape.

Slip the hanging loop over a doorknob or tape it temporarily to a table edge so that the mobile can hang freely. Help children adjust the balance by winding the short thread toward the center of the straw. Quickly tape the thread to the straw in its balanced position, as shown in Figure 12–2. Tape completed mobiles to a ceiling beam or a doorframe.

Dramatic Play

Store. Put a pan balance or old kitchen scales, containers of materials like horse chestnuts, and artificial fruit and vegetables in the play store area.

Block Play. Help block builders think about why their structures collapse. Suggest making broad, sturdy bases for tall towers, so that gravity will pull more on the bottom of the tower than on the top. When appropriate, ask if gravity is pulling equally on both sides of the weight-supporting blocks or whether an unbalanced load will tip over.

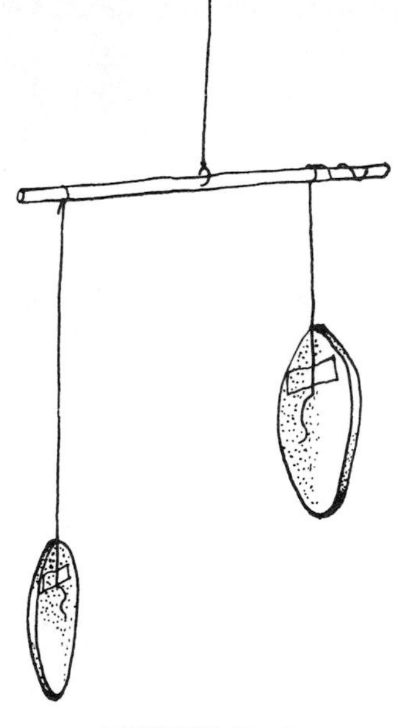

FIGURE 12–2

Spaceship. Try to find a refrigerator shipping carton for the children to convert into a spaceship. Help them improvise space helmets with small cartons and coils of telephone cable wire like the one worn by Little Bear on his imaginary trip to the moon. Before setting up the spaceship play, supply play ideas by reading *Little Bear,* by Else Minarik or *Curious George Gets A Medal,* by A. A. Rey.

Food Experiences

Take the class to the nearest food store where purchases are weighed on a balance beam scale. Buying and sharing a half pound of peanuts with the children is a good investment in gravity learning.

Thinking Games: Logical Thinking and Imagining

Use gravity pull as a topic to stimulate creative thinking. Ask, "What if there were no gravity pull from Earth? What would be different? What would it be like to play on the playground? What if we had to eat our meals as the astronauts do on spaceflights? Would we be sitting in our room?"

Field Trips

Knowing how much gravity pulls on objects is an important part of many businesses and services. Try to include a look at weighing devices during the field trips you plan to the post office, grocery store, feed store, airport, drugstore,

All of us together weighed a lot on the loading platform scale.

medical offices, or the loading docks of factories. It can be great fun for a whole class of children to be weighed together on loading dock scales.

Creative Movement

Help the children recognize the way their bodies involuntarily adjust to changes in body position in order to maintain their balance when they dance or exercise. When they lean in one direction, their bodies automatically compensate for gravity's pull by extending an arm or leg in the opposite direction.

Let the children pretend that they are on a ship in a storm, rocking from side to side. They will stretch their legs into a wide-based stance or tip over. Suggest moving like ice skaters, swinging arms and bending bodies as they glide and stride on the ice. Now they are tightrope walkers sliding one foot in front of the other on a swaying rope, arms extended. Next, imagine that they are going to lift a heavy rock, working very hard with their muscles to force it up against gravity's pull, bit by bit. Now put the rock down carefully and push it ahead slowly, then move like the astronauts on the moon where the gravity pull is weak.

In *Creative Movement for the Developing Child,* Clare Cherry suggests two poems as stimuli for balancing movements: "The Cat on the Fence" and "The High Wire Walker."

Whenever tension seems to be mounting in the classroom, or when excited children need help shifting from physical activity to a learning situation that requires concentration, use this "anti-gravity relaxation fantasy." Guide it slowly and quietly, saying: "Let your eyelids slide down now to close . . . and let your hands be loose and comfortable in your lap . . . Let your arms be loose . . . and limp . . . and light . . . so light they seem to float . . . so light that it seems that gravity isn't pulling on them much at all . . . And now . . . notice . . . just notice how loose and light your neck and shoulders feel . . . There is just enough gravity pull to help them keep your head above them . . . But your head might want to move alllll around slowly, till it finds a comfortable place to rest. There, it feels soooo comfortable now . . . There is almost no heaviness in your body now . . . Everything feels easy now . . . Let your mind take you on a floating journey now . . . You might want to have a cloud to rest on, feeling all that softness around you . . . or you might want to be an astronaut floating inside a space capsule, far, far, away from Earth's gravity pull . . . It's your floating journey, so you can have it be just the way you want it to be, because gravity doesn't pull on your thoughts—ever . . . so you can float as easily as you like in your imagination . . . And then gradually, slowly, you begin to notice gravity's pull again . . . You notice how well you are breathing now as you drift down closer, closer to Earth . . . Feeling rested and fresh and relaxed, now . . . And you take three long, slow, deep breaths . . . and then your eyes open again, so that you can see what we will be doing next in school."

SECONDARY INTEGRATION

Keeping Concepts Alive

One child applied gravity concepts in a judgmental way when he picked himself up from a fall complaining, "That ol' grabbity pulled me down!" Many of the large muscle-play activities provide opportunities to point out how gravity's pull makes some of our fun or work easier, such as balancing block towers, enjoying a slide, a teeter-totter, or swing, playing catch, or pouring water from a pitcher. Some activities are difficult to do because of gravity's pull. This is why children become tired when they put away blocks, climb the jungle gym, walk up the stairs, or trudge up a hill.

Relating New Concepts to Existing Concepts

Gravity/Air Relationship. Balancing two air-filled balloons is a vivid way to demonstrate that gravity pulls on air. To make a long balance arm, insert one drinking straw partway into another straw. Insert a hanging loop in the center as you did in making the mobile. Blow up two balloons to equal size; tie one to each end of the arm. Be sure that the balloons are the same color. Some children may believe that balloons of different colors are also different in the amount of air they contain.

Hang up the arrangement and let the arm come to rest. Add bits of masking tape to the end of the arm that is higher until balance is achieved. It is hard to blow exactly the same amount of air into two balloons. Explain to the children that you are going to prick one balloon. "What will happen to that balloon?

Some children become scared when the balloon pops and forget to watch what happens to the balance. Put your hands over your ears if the noise might bother you." Pop the balloon. "Which side of the balance went down? Why did that happen? Gravity pulls on air, too!" (If the pricked balloon flies apart, gather the pieces and tape them onto the end of the balance stick so a comparison of empty and air-filled balloons may still be made.)

Gravity/Water Relationship. Introduce the idea that gravity's pull plays a part in determining which objects sink and which float. Gravity pulls on water more than it pulls on corks, for instance. Recall that raindrops are pulled down to Earth when they become too heavy to float as droplets of vapor in clouds (see chapter 9, Weather and Seasons).

Family Involvement

Families often have opportunities to show their children things that go up and away from gravity's pull. Big motors help elevators and escalators lift people. Jet propulsion or propellers help lift airplanes off the ground. Heavy motors are needed to lift loads of materials at building construction sites which they could visit together.

EXPLORING AT HOME*

A Balky Balloon. Have some fun with gravity! Blow up two small balloons of the same size and color, if possible. Tie one; let the air out of the other. Using a small funnel, put a tablespoonful of uncooked rice or a few small pebbles into the emptied balloon. Now blow it up again, and tie it. Let your child try to blow the two balloons over the edge of a table. Which one rolls off? Which one tips, but stays put on the edge of the table? Can your child figure out how gravity made the difference? Compare the weight of each balloon to find out.

A Balancing Act. You'll need an old-fashioned peg clothespin, a foot of lightweight wire, and two old keys or metal washers of about the same weight to make a neat gravity toy. First, see if you or your child can balance the clothespin upside down on the tip of one finger, or on the palm of your hand. Then, twist the center of the wire around the neck of the upside-down clothespin. Twist each end of the wire through the loop of a key or the hole of a washer. Pull these ends of the wire downward into an arched curve, so the keys or washers hang down lower than the clothespin. Adjust the position of the keys or washers so the weight is equal on both sides of the clothespin. Now try the balancing act! Can your child tell you how gravity makes this possible? (Teachers sending this message to families might add a simple sketch, referring to Figure 12–1.) Gravity pulls more to each side and below the clothespin, so it can stand straight up on its head this way.

*These suggestions may be duplicated and sent home to families.

If you want to do more gravity experiments with your child, you'll find more ideas in *More Science Secrets,* by Judith Conaway.

RESOURCES

AMERY, HEATHER. *The Know How Book of Experiments.* Saint Paul: EMC Publishing, 1978.
MILGROM, HARRY. *Adventures with a Cardboard Tube.* New York: E. P. Dutton, 1972.
WILLIAMS, R., ROCKWELL, R., & SHERWOOD, E. *Mudpies to Magnets.* Mt. Rainier, MD: Gryphon House, 1987. Includes directions for making bleach bottle "space helmets" and for making a balance toy.

13

Simple Machines

Throbbing motors and turning wheels make little-noted daily background noises in the Western world. Children are proud to learn how to move and lift things with the help of simple machines. The knowledge invites interest in the complex machines they have previously taken for granted. The learning experiences in this chapter explore the following concepts:

- Effects of friction: heating, slowing, and wearing away objects
- A lever helps lift objects
- A ramp shares the work of lifting
- A screw is a curved ramp
- Simple machines help move things along
- Some wheels turn alone; some turn together
- Single wheels can turn other wheels
- Single wheels can help us pull down to lift up

Experiencing the advantages and disadvantages of friction is a useful preliminary to the experiments that follow. The simple machines experiments illustrate the lever, ramp, screw, wheel and axle, and pulley.

Introducing Friction. "Rub the palms of your hands together as fast as you can. Keep going. Do you feel something happening to your hands? What we are doing makes heat. The reason for the heat is called *friction*."

CONCEPT: Effects of friction: Heating, slowing, and wearing away objects

1. Does friction slow sliding objects?

LEARNING OBJECTIVE: To experience firsthand the slowing effect of friction on a slide.

MATERIALS:

Gum erasers

Small pieces of waxed paper

Magnifying glass

Indoor slide or teeter-totter plank, propped against a table

Sheets of waxed paper

Rubber sink or tub mat

LARGE GROUP ACTIVITY:

1. Pass around pieces of waxed paper. "Try rubbing this on your arm. Does it slide fast? Try sliding the rubber eraser. Does it slide fast on your arm? Can you think of a reason for this?" (The rubber makes more friction.)
2. "Look at paper and the eraser through the magnifying glass. Which is smooth? Which is fuzzy?"
3. Let children take turns on the slide, sitting first on waxed paper, then the rubber mat. "Smooth paper doesn't make much friction, so you move fast."

2. Does friction make heat and wear objects away?

LEARNING OBJECTIVE: To experience firsthand the heat produced by friction and to actively wear away materials by producing friction.

MATERIALS:

6" (15 cm) pieces of scrap lumber

Coarse sandpaper, cut into 3" (7 cm) squares

Hammer

Nails

Magnifying glass

GETTING READY:

Hammer a few nails half their length into a chunk of wood. (To use the hammer as a lever, catch nail in claw and roll hammer back on curve of claw. Pull down on handle to lift nail up and out.)

SMALL GROUP ACTIVITY:

1. Give each child sandpaper and scrap lumber. "Feel these. Are they smooth or rough? Do they look smooth or bumpy with the magnifier? Rub them together hard and long. See what happens."
2. "Are your fingers feeling warm? (Rough things rubbing together make lots of friction.) Are bits of wood wearing away? Look at your wood. Feel it. Is it getting smoother? Is the sandpaper changing? What does friction seem to do to things?"
3. "I'm going to pull a nail from this wood. Do you think that will make friction? Let's feel the nail quickly after it comes out. Is it warm or cool? Do you think the nail and wood rubbed together? How do you know?"

3. Can we cut down friction?

LEARNING OBJECTIVE: To notice how lubricating materials reduces friction.

MATERIALS:

Coarse sandpaper

Petroleum jelly

Meat trays

Peanut butter

Soda crackers

Waxed paper

Table knives

SMALL GROUP ACTIVITY:

1. Give children two pieces of sandpaper. "Rub these together fast. What happens to the bits of sand?"
2. "Look at the sandpaper. Is it bumpy and rough? Do you think it would be smoother if something filled up the bumpy places? Try this petroleum jelly on it. See if the pieces of sandpaper slide past each other smoothly now."
3. Give each child two crackers on a tray. Repeat steps 1 and 2, offering peanut butter as a lubricant. Let the children eat this experiment.

GETTING READY:

Cut sandpaper into 2″ (5 cm) squares.

Put a spoonful of peanut butter on pieces of waxed paper for each child.

Group Discussion: Show the children a small can of household oil. Ask what they know about it and why it is used at home. Develop the idea that oil makes a smooth sliding surface on objects that rub together. It cuts down friction that makes heat, wears bits away, makes things difficult to use, and makes things squeak.

4. Can friction be useful?

LEARNING OBJECTIVE: To notice that friction is needed for writing, drawing, and gripping things.

MATERIALS:

Chalk

Dark-colored paper

Small pieces of waxed paper

Pencils

Petroleum jelly

SMALL GROUP ACTIVITY:

A

1. Give children paper and chalk. "What sound does the chalk make on the paper? Scraping? Scratching? Why do you think this happens?"
2. "Try dipping your chalk into the petroleum jelly. Now does it rub and scrape? What happened to the chalk mark? Do you need friction to draw?"
3. Let them try to draw on waxed paper with pencils. Explain that the wax coating is like a lubricant.

MATERIALS:

Screw-top plastic containers, such as small paste jars

Two pans of water

Soap

Towels

B

1. "Are your hands strong enough to unscrew this jar cover? You can do it well."
2. Try it again, but this time get your hands wet and soapy first. Why is it hard to do now? What is missing?"
3. "Try the doorknob with wet hands. Can you turn it?"

Group Discussion: Ask the children to check each other's shoe soles. Are some smooth and slippery? Do some have rubber heels? Do others have one-piece rubber bottoms? Are sneakers good for running and stopping, or do they skid? Are car tires smooth and slippery, or more like the soles of sneakers? Why? Talk about slippery road conditions, slippery bathtubs, and wet bathroom floors. How does friction make these places safer?

CONCEPT: A lever helps lift objects

1. What can a lever do?

LEARNING OBJECTIVE: To discover how to arrange a plank and resting point as a lever to lift and be lifted.

FIGURE 13–1

Introduction: "One day I watched Amanda on the playground pushing *down* to lift David *up* in the air. Then David pushed *down* to lift Amanda *up*. What were they doing? (Using a seesaw.) Could Amanda lift me that way? Gravity pulls much more on me than on her. Let's see if we can find a way."

MATERIALS:

Six-foot seesaw plank

Block, books

12" rulers or metersticks

Pencils

String

Hammer, nails

Scrap lumber

Sturdy box large enough to hold two seated children

SMALL GROUP ACTIVITY:

1. Put a block below midpoint of the plank. Have a child stand on one end. Ask another to stand on the raised end to lift the other.
2. "Could one of you lift me up that way? Try. Now let's see what will happen if we move the resting-point block closer to me. Now can you lift me? Yes! We changed the resting point and made a lever. It helped her lift me *up* when she pushed her weight *down*."
3. To show the importance of the resting point, place the box on one end of the plank. Put the block nearby. Ask two children to sit in the box. "How could we use the plank to lift this load?" The load tips if the *plank is lifted up*. The load is *lifted up* as the plank is *pushed down* only when the plank and block are used together as a lever and a resting point.

Children who are waiting for turns to lift the box with the plank lever can try out two other levers. Tie two or more books together with a string. Loop one end of the string so children can try to lift the bundle with one finger. Offer a ruler and a pencil for a resting point to make a lever. "It was hard to lift up the bundle pulling *up* with one finger. Try pushing *down* on the ruler lever to lift the books up." (Figure 13–1).

Drive nails into the wood about half the nail length. "It can be very hard work to pull a nail out by pulling up. Try using the hammer as a lever, pushing down to lift the nail up."

Group Discussion: Look at the picture of a caveman using a branch and a small rock to move a huge rock in *Simple Machines* by Rae Bains. Pry open an empty cocoa can with the handle of a spoon. Ask if the spoon was a lever. Show that the resting point was the edge of the can. Talk about oars on a rowboat serving as levers that are pulled *back* to push the boat *forward*. Sing "Row, Row, Row Your Boat."

CONCEPT: A ramp shares the work of lifting

The inclined plane (ramp) is another simple machine that helps us lift things against gravity's pull.

1. Can a ramp help us lift?

LEARNING OBJECTIVE: To find out firsthand if a ramp helps lift a heavy load.

MATERIALS:

Old, sturdy suitcase

Heavy objects to fill suitcase

Plank, at least 4′ (1.2 m) long

Table

GETTING READY:

Fill the suitcase with enough weight to make it hard for a child to lift by the handle.

SMALL GROUP ACTIVITY:

1. "See if you can pick up this heavy case by the handle and lift it onto the table."
2. "See if this plank can make it easier to get the case up on the table." Lean the plank from the floor to the tabletop. Place the suitcase near the bottom of the plank. "We call this a ramp."
3. "A ramp holds some of the weight of the suitcase as you slide it along. Now you can get the suitcase up there."

Group Experience: Make a miniature steep mountain of damp sand in a long cake pan or in the sand table if you have one. Use a tiny matchbox car as a demonstrator. "Let's see if a car can go straight up the side of a steep mountain like this. What will happen if I let go? Gravity pulls it down *fast.* Could we curve a ramp road around this mountain to help the car go up?" Make a road like this, then let children try the car on the road. What happens when the children let go of it? (It may stand still in the damp sand, but on a real mountain road, gravity could slowly pull a car down.)

CONCEPT: A screw is a curved ramp

Introduction: Draw a thick crayon line diagonally from one corner of a sheet of typing paper to the other. Cut along the line, leaving a crayon-edged triangle. Bring the triangle, a pencil, a large screw, and a collection of nuts and bolts to a class discussion. "Does the line on this paper triangle look like a ramp? Let's see what we can do with it." Wind the triangle around the pencil so the line looks like a screw thread. Trace the spiral path from the bottom of the ramp to the top. Compare it with a screw. Give children the bolts and nuts. "Can you make the nut go up the ramp that curves around the bolt? Try it."

1. Can screws lift objects?

LEARNING OBJECTIVE: To investigate how threaded objects move up the corresponding threads of a screw.

MATERIALS:

(Use as many as possible)

Screw-type cookie press

Tissue

C-clamps

Modeling clay

Old lipstick tube

Screw-type nut cracker

Screw-top plastic jars

SMALL GROUP ACTIVITY:

1. Remove design plate from the cookie press. Stuff a tissue in the press. Let children see what happens when the screw knob is turned. "What lifted up the tissue?"
2. Put a small ball of clay on the end of the C-clamp, turning the screw. Hold clamp upright; turn the screw until the clay rides up to be squashed by the top.
3. Show how two screws can be threaded together so they can't be pulled apart (use jars and pipes).

Plumbing pipes and joints (now available as plastic toys)

Old piano stool or office swivel chair

GETTING READY:

The bottom half of a lipstick tube is usually two pieces. Separate them to reveal the spiral ramp that the lipstick moves along as it is twisted up. Show children how this device works.

4. Let children enjoy working with the small materials you have assembled before presenting the experience of being lifted by a screw on a piano stool or office chair. Take time to examine the emerging screw as it turns.

CONCEPT: Simple machines help move things along

1. How do rollers move things?

LEARNING OBJECTIVE: To experience and compare the difference between dragging a box, pushing it over rollers, and pushing it with wheels under it.

MATERIALS:

Grocery cartons or nested hardwood crates*

Jump ropes

4 cut-off broomsticks or cardboard cores from newsprint rolls to use as rollers

Platform dolly or board and 4 caster wheels

GETTING READY:

If the school custodian doesn't have a platform dolly for scrub buckets make your own by screwing 4 swivel casters into a 1" (2.5 cm) thick plank.

SMALL GROUP ACTIVITY:

1. Loop jump ropes around cartons. Let children take turns pulling each other in boxes, using jump rope handles in each hand.(Cardboard boxes fall apart soon, so have many.) "Is pulling easy or hard?"
2. Place rollers side by side on the floor. Put box on top. Repeat 1. "See if rollers help move the box. (They don't stay under the box.) Put them back for the next child's turn."
3. "Which way made less friction?" Feel box bottoms after the dragging and after rolling.
4. "Let's try rolling wheels that will stay under the box now." Put box on the platform dolly. "Which is the easiest way to pull the box: dragging, with rollers, or with wheels?"

*The nested hardwood crates have metal glides on the bottom to reduce friction. The largest crate has swivel casters. These crates are available as play equipment from various toy manufacturers.

Group Discussion: Recall the rollers experiment. "Was the carton moved very far by four rollers? Would rollers be a good way to move cars and trucks? What if someone had to keep putting rollers in front of the car to keep it moving? Rollers are used that way when whole houses are moved a short distance. They give good support to the house. Rollers are also used to unload big trucks at the grocery store. Rollers are part of conveyor belts and escalators."

CONCEPT: Some wheels turn alone; some turn together

1. *How do single wheels and pairs of wheels work?*

LEARNING OBJECTIVE: To notice the difference between things that move on single wheels and those that move on pairs of wheels joined by axles.

MATERIALS:

A light piece of furniture on 4 swivel wheels (office chair, typewriter stand, utility table, crate on casters, platform dolly)

Toy car with axles exposed

Small paper plates

Masking tape

Plastic drinking straws

Compass

GETTING READY:

Find the exact centers of paper plates with a compass. Mark each center; push out a $\frac{1}{4}$'' (1 cm) hole with pencil tip. Have one for each child.

Cut 1'' (2.5 cm) pieces of tape; stick lightly on a nearby table edge for children's use.

SMALL GROUP ACTIVITY:

1. Divide children into two facing rows, at least 5 feet apart.
2. "Let's take turns pushing the chair to children across the room. Watch what happens. How does it move?"
3. "Now let's push these cars across. Do they move like the chair wheels?" Compare with the swivel again, watching the wheels very closely. Turn the swivel and the cars upside down to see if they look the same. Supply the word *axle* if a child doesn't offer it. Point out pairs of wheels that turn together on an axle.
4. Let children try to roll their plates across to another child. Offer straws and tape, suggesting that two children could join wheel plates to see if they will roll better as a pair with an axle.

Group Discussion: Talk about the advantages of single wheels and pairs of wheels. If your classroom has an old upright piano or a classroom storage chest on casters, push it with the children. Tiny, single wheels can make it easier to push something very heavy for a short way.

"Do you think wagons, cars, buses, and trucks have a single wheel on each corner?" Show some pictures of wheeled objects cut from catalogs. "Do these things use single wheels or pairs of wheels on axles?" Children are surprised to learn that doorknobs are pairs of wheels on an axle. Bring in an old doorknob or unscrew one at school. Look together at a manual typewriter for rollers, wheels, and axles.

CONCEPT: Single wheels can turn other wheels

1. *How do gears work?*

LEARNING OBJECTIVE: To observe how sets of gears mesh to operate mechanisms.

MATERIALS:

Eggbeater (hand operated)

2 deep mixing bowls

Soupspoons

Water

Detergent

Gear toys, visible clock, visible music box works, old clock to open

Crayon

Newspapers

Sponge

GETTING READY:

Put $\frac{1}{2}$ cup water and 1 teaspoon detergent in each bowl.

Spread papers on the floor.

SMALL GROUP ACTIVITY:

1. "Let's look very closely at a beater. How many wheels do you see? (Some wheels may look different.) Watch what happens to the little wheel when the big wheel is turned." Introduce the term *gear,* explaining that this is a wheel with teeth that mesh with teeth of the wheels next to it.
2. "Let's take turns using the beater to find out how the gears work. Let's compare beating bowls of soap suds with the beater and with a spoon to find out which works more easily."
3. Help children see that the big gear turns the little gears on the beater blades much faster.
4. Make other gear toys available for independent inspection.

Note: Children enjoy taking an old windup clock apart, using a screwdriver. However, the coiled steel spring is *sharp.* It should be unscrewed and removed only by an adult.

Group Discussion: Try to borrow a one-speed bicycle to examine with the group. Turn it upside down and look at the gears. "Sometimes gears turn each other without touching each other. Instead they fit into a chain that moves them. Is one gear large and one small, like the eggbeater? Let's turn the pedals and watch the chain make the small gear turn. Count how many times the back wheel turns while I turn the pedal one time." Count one turn when the air valve reaches the top.

Look at the tire tread. "Could friction help stop a bicycle? Why do you think the chain looks greasy? Would it need grease? What do you think is inside the tires?"

CONCEPT: Single wheels can help us pull down to lift up

1. Does a pulley make lifting easier?

LEARNING OBJECTIVE: To experience by comparison how a pulley makes lifting easier.

MATERIALS:

Small pulley

Firm cord, double the length of floor-to-pulley distance

Small school chair

Stopwatch

Screw hook

SMALL GROUP ACTIVITY:

1. Show children the pulley. "Do you suppose this single wheel could make it easier to lift a chair?"
2. "First let's see how long you can hold this chair with one hand. Lift it as high as you can. I'll time you." Record length of time each child can hold chair.

Two proud boys, a rope, and a pulley lift and hold a school chair at ceiling level.

GETTING READY:

Install screw hook in a ceiling beam or door frame (ask custodian).

Pass the cord over the pulley wheel; hang pulley on screw hook.

3. Tie cord firmly to chair back. "Now try pulling down on this cord to lift the chair. I'll time you to see how long you can hold it up." (Demonstrate careful lowering of chair, but stand by in case someone lets it down too fast.) "You are pulling down on the rope to lift the chair up. Does the pulley make the lifting job easier?"

Group Discussion: (Try to find pictures of the following things to show.) "Pulleys help people load big ships, trains, and barns. Pulleys are used on steam shovels, on sailboats, on flagpoles, on scaffoldings, in draw-drapery rods, and inside the casings of windows. A pulley arrangement moves people up in large buildings on elevators and escalators. Single wheels help people and machines pull *down* to *lift up* or *pull* across." Richard Scarry's book, *What Do People Do All Day?* is an excellent source for pulley illustrations.

PRIMARY INTEGRATING ACTIVITIES

Music (Resources in Appendix 1)

1. Sing this song about friction to the tune of "Here We Go Round the Mulberry Bush" while rubbing hands together.

Friction Song

This is the way we warm our hands, warm our hands, warm our hands.
This is the way we warm our hands, out in chilly weather.

Friction is what warms our hands, warms our hands, warms our hands.
Friction is what warms our hands, rubbing them together.

2. Beatrice Landeck's song "Little Red Wagon" is a lively reminder of the usefulness of wheels and axles. One child created these three verses for his favorite song:

   "Both wheels are off and back end's dragging . . ."
   "Four wheels are off and we're not even moving . . ."
   "Now we're going to the repair shop . . ."

Math Experiences

Forming Sets. Use plastic cars (sold in bags at variety stores) and macaroni wheels as members of sets to link simple machine study with math activities.

Car Classification Game. Let children park plastic cars in the correct parking area according to color. Use red, yellow, and blue construction paper as the parking garages where the cars of each corresponding color are parked. Add a second dimension to the classification so that red cars with tops are parked on one side of the red garage, red cars without tops are parked on the other side of the red garage, and so on.

Wheel Quantity Classification Game. Cut out and mount catalog pictures of wheeled objects. Ask children to sort them into groups according to the number of wheels each object has. (A record player turntable could be used as an example of a one-wheeled object.)

Stories and Resources for Children

ARDIZZONE, EDWARD. *Little Tim and the Brave Sea Captain.* New York: Scholastic Book Services, 1968. The use of pulleys can be pointed out in lifeboat launching and in the rope rescue of Tim and the Captain.

BAINS, RAE. *Simple Machines.* Mahwah, NJ: Troll, 1985.*

BAKER, BETTY. *Worthington Botts and the Steam Machine.* New York: Macmillan, 1981. A good stimulus to creative thinking about machines children might wish to invent.

HORVATIC, ANNE. *Simple Machines.* New York: E. P. Dutton, 1989. Fine photographs illustrate familiar objects that use simple machine principles. The clear text is suitable for early readers and pre-reading listeners.*

REY, H.A. *Curious George Rides a Bike.* New York: Scholastic Services, 1973. A hammer is used as a lever to open a crate. Gears, chain, and pedals on George's bike, and pulleys on a circus wagon cage are shown in this adventure.

*Starred references, written at the young child's level of understanding, can help teachers with minimal backgrounds in science expand their knowledge base.

ROCKWELL, ANNE & HARLOW. *Machines.* New York: Macmillan, 1972. Beautiful watercolors illustrate simple machines. The text is direct, brief, and very appropriate for young children.

SCARRY, RICHARD. *What Do People Do All Day?* New York: Random House, 1968. A profusion of funny illustrations show wheels, gears, pulleys, rollers, and screws in use.

WYLER, ROSE. *Science Fun with Toy Cars and Trucks.* New York: Simon & Schuster, 1988. Simple text for young readers. Toys are the basic equipment for experiments with ramps, friction, wheels, and the laws of motion. An important experiment demonstrates the purpose of seat belt use. Paperback.

ZUBROWSKI, BERNIE. *Wheels at Work.* New York: William Morrow, 1986. This Boston Children's Museum activity book guides the building of simple wheel devices with inexpensive materials.

Poems (Resources in Appendix 1)

To stimulate children to invent machines they may wish for, read "Homework Machine" from *A Light in the Attic,* by Shel Silverstein.

Fingerplays

The familiar fingerplay, "The Wheels of the Bus," calls for arm rolling and active bouncing.

Art Activities

Collage. These collage materials are strongly suggestive of simple machines: macaroni wheels; pieces of thick string; bottle caps; Popsicle sticks; and round, triangular, and rectangular construction paper shapes. (The macaroni wheels may become essential parts of pulleys and gears, or they may be quietly eaten by a child whose inspiration is not as strong as his curiosity about new tastes.)

Cutting and Pasting. Young children enjoy making their own books or contributing pages to a group book for classroom use. For individual books, staple a few sheets of folded paper together. For group books, use cardboard punched with holes at top. Fasten with leather thongs or notebook rings.

Provide scissors, paste, and pages cut from catalogs illustrating things with simple machine parts: clocks, eggbeaters, wheelbarrows, wheeled toys, mechanical toys, tools, carts, pepper mills, and so forth. If the supply of pictures is sufficient, they could be part of a math experience, making separate pages to show one-wheeled objects, two-wheeled objects, and so on.

Friction as an Art Medium. Sometimes when children are drawing with pencils and crayons, or making chalk rubbings, remind them that friction is involved in making colors stay on the paper.

A Friction Project. This project requires at least two days. Let the children sand 3" x 5" (8 cm x 12 cm) pieces of scrap lumber to satin smoothness. Next, draw a design on paper with magic markers to cut out and glue to the sanded wood. The teacher can coat the plaque and picture with clear varnish and allow at least a day of drying time. Insert a small screw eye in the top edge of the plaque for hanging. This makes an appreciated gift for parents.

Dramatic Play

Packing Carton Vehicles. Children usually need little more than grocery cartons, paper plate wheels, and paper fasteners to create play cars and trains. A real steering wheel from an auto salvage yard, an inner tube and tire pump, and some sets of discarded keys add fun to the play. Books like *Behind the Wheel,* by Edward Koren, can supply play ideas.

Community Playthings sells a fine demountable, two-seat car for children to assemble and use. Initial adult guidance on connecting wheels and axles will be needed. Completing the car requires teamwork and problem-solving ability.

Elevator. A tall refrigerator packing carton can become an elevator with a door cut in one side and numerals marked above it to indicate floors. It can be part of block or housekeeping play as desired.

Pulley Fun. Make an elevator for a skyscraper. Ask children to build a three-sided block tower directly below a pulley. Fasten a milk carton elevator to one end of the pulley cord. Cut a door in one side of the carton and fill it with toy passengers.

Attach two small pulleys to opposite sides of the block play area, a few feet above the floor. Tie the handle of a small basket to the pulley cord and knot the cord into a continuous, taut loop running between the two pulleys. It can be an aerial tramway for toy passengers or a conveyor belt for block construction workers.

Other Indoor Fun. A toy conveyor belt ramp is sold commercially. It can be an airplane baggage loader or a piece of play farm equipment. The two rollers that move the belt are operated with a small crank.

Add a toy sand or water wheel to sandbox or water play arrangements. They are sold seasonally as beach toys.

Provide pieces of perforated hardboard and hardware odds and ends to attach to it: cupboard doorknobs and hinges, nuts, bolts, and washers. Short screwdrivers are easiest for children to control.

Many commercial toys with visible working parts extend the simple machine concepts in play. Examples include construction sets (such as Tinkertoys), gear sets, toy clocks, music boxes, and locks with visible working parts. A sturdy "look-inside" music box and a pulley-operated cable car kit are both available by mail from Hearth Song (see Appendix 4).

A broken hand-wound alarm clock is a fine source of firsthand information about gears and springs. Provide small screwdrivers such as those sold as sewing machine accessories, and let the children take apart the clock. Springs can be sharp, so adult supervision is required for the dismantling.

Supply the workbench with bottle caps, metal ends from food container tubes, spools, and empty typewriter ribbon reels to inspire the construction of wheel-, gear-, and dial-encrusted inventions.

Food Experiences

Children feel very important when they turn a crank that contributes to good eating. Several simple machines can be used to prepare food in the classroom. Make molded cookies with a cookie press. Use a hand-cranked grinder to make graham cracker crumbs for unbaked cookies or to make toast crumbs for the birds. Use a rolling pin roller to make cut-out cookies and a gear-driven egg-beater to mix instant pudding.

Thinking Games: Imagining

Inventions. Read Shel Silverstein's poem "Homework Machine." Then ask children to imagine a new machine that would help them in some way. "What would it do? What would it look like?"

What If? Collect pictures to illustrate a family picnic outing: getting food ready, climbing into the car, and gathering fishpoles and playthings. Ask, "What if there weren't any friction anywhere one day? What would be different for this family on a picnic? Yes, the mother would have trouble opening the peanut butter jar to make the sandwiches, the children would skid and fall down trying to walk to the car, the baseball bat would slip out of the boy's hand when he hit the ball, the fishpoles might slide out of their hands."

Field Trips

1. Field trips to see simple machines in action can be as simple as a walk to the school kitchen. A well-timed visit might allow the children to see a food order being unloaded from a delivery truck on a hand truck. Hand trucks have one pair of wheels on an axle and use the lever principle to lift loads. The kitchen may have a swivel-wheeled utility cart and a large can opener that has a sharp wheel to cut metal, gears to turn the can, a crank that turns an axle, and a screw to clamp the opener to a table.
2. The school office may have an electronic typewriter that has wheels, gears, axles, and a roller inside. Ask permission to lift the top for children's visual inspection (no touching!). Examine the gears of the plastic correction tape reels and the ribbon cartridge. Find the corresponding gears inside the type-writer that mesh with and turn them.
3. A repairman working in the building could show children an array of tools and equipment that apply simple machine principles.
4. The nearest driveway could be the scene of an impressive sight if the teacher can demonstrate the screw or lever principle by jacking up her car. (Be sure to observe the safety precaution of blocking the wheels on the ground with a brick.)

Be sure to check for safety hazards when planning the following field trips. The locations are not intended for general public use. Children should be well supervised.

5. A visit to an auto repair shop can verify the usefulness of simple machine principles. Mechanics in small garages may use scooting platforms on caster wheels to get under cars. A mechanic might demonstrate the tire changer, using a crowbar as a lever to pry off the tire.
6. Many wheels, rollers, conveyor belts, and pulleys are used in the work areas of the post office. Each one also has a flag pole with a pulley to raise the flag.
7. Backstage supermarket operations include hanging meat from overhead rolling tracks, using sets of rollers enclosed in metal frames to slide cases of food from trucks to storerooms, and moving groceries on checkout counter conveyor belts.
8. An older retail store in your community might still use a hand-operated dumbwaiter to bring stock upstairs from a basement storage room. In some types of dumbwaiters, the pulleys and ropes are visible. See if there is an enclosed freight elevator near your school where the whole class could take a ride together. It's fun to do.

There are so many opportunities to see simple machine principles in action that it would be easy to overdo the effort. Don't risk wearing down children's interest by taking too many field trips.

Creative Movement

Mechanical Toys. Children enjoy acting out the stiff, jerking movements of windup toys. Offer to wind up one child toy at a time so that other children can guess what kind of toy they are watching, or wind up all the toys at once so they can all move to music. In addition, children can try to move like the Tin Woodsman, both when he was too rusty to move and after friction was cut down by oil so that he could move.

SECONDARY INTEGRATION

Keeping Concepts Alive

There are countless opportunities to bring simple machine principles to light during everyday school activities. When we want things to move or spin or slide easily, we can mention the need to cut down friction with a few drops of oil. There are times when too little friction poses safety hazards for young children: shoeless children moving fast on slippery floors or unwary children crossing streets on rainy or icy days.

For a vivid application of the lever and a pair of wheels, let children help trundle heavy sandbags to the sandbox with a hand truck. Encourage them to try moving the sandbags without the hand truck so that they will realize how much the hand truck helps them.

Simple machine concepts are part of workbench use. Examples include the friction of sanding and sawing, the lever action of pulling out nails with the hammer, and the interlocking screw threads that hold wood tightly in a vice or clamp.

Whenever possible let children watch the repair and maintenance of classroom and playground equipment. Better still, enlist the help of a shy child or one who needs a chance to get positive attention and approval. One child who broke toys and mistreated other children to attract attention was able to give up these undesirable behaviors when he became our chief mechanic. He took such pride in manning the oilcan and tightening tricycle bolts that he became a good steward of school property and a friend to other children.

Relating New Concepts to Existing Concepts

Simple machines help overcome the effect of gravity's pull. This can be brought out by speaking about the weight of the objects being lifted or moved in terms of gravity's pull. The effect of magnetism could be used to discover what a strong axle or wheel frame is composed of or to learn what kind of hinge rusts and needs oiling when it works hard and squeaks from too much friction.

Simple machine principles are used to move boats on water. Easy examples are the levers called oars and the paddle wheels that power the swanboats seen in Robert McCloskey's book, *Make Way For Ducklings.*

Family Involvement

Ask families if they would be willing to lend equipment or demonstrate skills that involve simple machines. Todd's mother loaned us a butter churn that her grandmother had used. Jenny's mother made bread with us, using her hand-cranked kneading bucket. Both children gained new status as a result of their parents' contributions.

EXPLORING AT HOME*

What and How? "What's inside?" "How does it work?" These are very familiar questions your child asks. Now he or she may be able to figure out some of the answers about the tools and devices you have in your house. Find out together by taking a close look at some of the simple machines you use:

*These suggestions may be duplicated to send home. Remind parents that they are not homework assignments. They are meant for having fun with the child and for helping the child realize that science plays an important part in life.

- A screwdriver becomes a lever when you pry open a paint can with it.
- Traverse drapery rods or certain kinds of blinds have small pulleys inside the top rod. Typewriters and some exercise equipment use pulleys, too.
- Lift the hood of your car to reveal the belts that let various shafts in the motor turn together, then explore the system with your child, discussing how each belt works.
- Gears are wheels that turn other wheels. Can openers have small gears, so do hand-operated egg beaters. Hand-wound clocks are full of gears. If you have one that's no longer being used, try to take it apart with your child, using the smallest screwdrivers you can find. (You should be the one to remove the flat, coiled spring. It may really spring up fast, and its edges are sharp.)
- Explore the workings of gears on bicycles of different speeds. Compare them to see how the chain and gears or sprocket wheels help the bike go faster with less effort than does a one-speed bike.
- Watch for simple machines like gears, pulleys and other wheels if you visit a science or historical museum with your child.

Easy Rollers. Check for small wheels that help you move things around the house. You may find them at the base of your refrigerator, vacuum cleaner, luggage or shopping cart, or outdoor grill. Take a close look at roller skates and skate boards. Make a game of counting all the wheels you see on trucks you pass on the road.

RESOURCES

MACAULAY, DAVID. *The Way Things Work: A Visual Guide to the World of Machines.* Boston: Houghton Mifflin, 1988. This book gives the teacher the same revelations of "so that's how it works," that children gain from their study of simple machines.

Teaching Resource

Moving Machines. Videotape. Bo Peep Productions, P. O. Box 982, Eureka, MT 59917. Parallels are shown between heavy construction machines at work and young children playing with toys like them.

14

Sound

The discovery that sound occurs when something vibrates can help children overcome fears about scary noises. Vibrations cause sound whether they originate in the distant clouds or in our own throats. The following concepts are explored in this chapter:

- Sounds are made when something vibrates
- Sound travels through many things
- Different sizes of vibrating objects make different sounds

The experiences begin with a group activity to clarify the term *vibration* and to establish the idea that vibrations cause sounds. This is followed by experiencing vibrations, experimenting with mediums through which sound travels, and relating the pitch of sound to the size of vibrating objects.

Introduction. Help children understand what a vibration is by producing one. Say something like this: "Can you lift your arm and let your hand dangle from your wrist? Now, shake your hand as fast as you can. What do you call what is happening to your hand? Shaking? Wiggling? Jiggling? Wobbling? Those are good words. *Vibration* is another word that means moving back and forth very fast. Does your hand look different when it is vibrating? Some vibrating things move back and forth so fast that they look blurry. Another thing happens when something vibrates. You can find out if you listen very quietly. Hold your hand near your ear, then vibrate your fingers and hand very fast. What happens? Does someone hear a soft, whirring sound?"

"Now put your fingers very lightly on the front of your throat, near the bottom. Very softly sing a sound like *eeeeee*. Do you feel something with your fingertips? Try it again. Did something inside your throat vibrate? Yes, you made a sound in there. Whenever we hear a sound, it is being made by something vibrating. Try touching your lips while you say *hummmmmm*. What's happening?"

CONCEPT: Sounds are made when something vibrates

1. Can we see things vibrate?

LEARNING OBJECTIVE: To learn the meaning of the term *vibration* by creating and watching vibration.

MATERIALS:

Binding strip from vinyl folder cover, or flexible ruler

Coffee can with plastic lid, or sturdy drum

Sand, rice, or foam packing chips

Quart milk carton

Rubber bands

GETTING READY:

Cut window opening in one side of milk carton.

Stretch rubber band around length of milk carton.

SMALL GROUP ACTIVITY:

1. "Watch and listen to this." Extend folder strip from edge of table or chair, like a diving board. Bend free end down, then release. "What did you see and hear?"
2. Show the milk carton. "What will happen if you pluck the rubber band? Watch and listen."
3. "It was easy to see the plastic strip and the rubber band vibrate. Some vibrations are hard to see. Let's put light things on this can lid to find out if the lid vibrates." Put sand, rice, or chips on lid. Tap the lid. What happens?

2. Can we feel things vibrate?

LEARNING OBJECTIVE: To learn the meaning of the term *vibration* by creating and feeling vibration.

MATERIALS:

One or more of these: windup alarm clock, kitchen timer, toy music box

Combs

Waxed paper (tissue paper disintegrates when damp)

GETTING READY:

Cut pieces of waxed paper to fold over teeth of combs.

SMALL GROUP ACTIVITY:

1. Pass unwound clock, timer, and music box among the children. "Do you feel these things vibrating now? Are they making sounds?"
2. Wind the equipment and pass again. "Now what do you feel and hear?"
3. "Try making some music with vibrating air and paper. Hold the paper at the bottom of the comb lightly between your lips. Now try to hum and blow a tune at the same time. The vibrations feel funny, but the sound is nice."
4. Tour the room and halls to feel vibrating electric motors (aquarium water pump, sound system speaker, water cooler, etc.)

Read *Whistle for Willie,* by Ezra Jack Keats, a story about how to pucker the lips tightly to blow out another kind of vibration.

CONCEPT: Sound travels through many things

1. Does air carry sound?

LEARNING OBJECTIVE: To become aware that sound vibrations can make air vibrate to carry sound.

MATERIALS:

12″ (30 cm) pieces of plastic garden hose

Plastic golf club cover tubes (very inexpensive at sports stores)

SMALL GROUP ACTIVITY:

1. "What could be inside the pieces of hose? Put one end next to your mouth; put the other next to your ear. Whisper your name into the hose. Could you hear yourself? What could be vibrating inside the tube?"
2. "You can *feel* what is vibrating if you put your hand at the end of the hose. Try saying words like *toot* and *tut*. Can you feel something pushing on your hand each time you say a word?" (Air carried vibrations from their throats through the tubes.)
3. Let seated children enjoy speaking to each other through the long golf club tubes.

Read *Goggles*, by Ezra Jack Keats.

Group Discussion: "Peek into the ear of the child sitting next to you. Do you see anything inside the ear that is vibrating? Could something invisible be in there?"

Recall how the sand or foam chips bounced around when someone tapped the coffee can lid. One vibrating thing made the things next to it vibrate. "Air vibrates when something next to it vibrates. That is the way sound is carried by air. It is the way sounds usually come to our ears. The air inside our ears also vibrates so we hear the sounds." Let children experiment with this idea by covering and uncovering their ears with their hands as they listen to some music. Could the air inside their ears vibrate very well when their hands were covering them?

Safety Note: "Our ears have an outside part that we can see and an inside part that we can't see. The inside part is so delicate that vibrating air can vibrate it. That is how we hear. We must be very careful with our ears to keep that delicate part working well. Can you think of some good safety rules for our ears?"

2. Does water carry sound?

LEARNING OBJECTIVE: To become aware that sound vibrations can make water vibrate to carry sound.

MATERIALS:

Two similar containers, such as plastic buckets

Pair of blunt scissors

Sponge, plastic sheet, or newspaper for spills

Golf club tube cut into 12″ (30 cm) lengths

GETTING READY:

Half fill one container with water

SMALL GROUP ACTIVITY:

1. "Take turns pressing one ear to the side of the bucket to listen to the sound I'll make." Open and shut scissors to make a steady clicking noise, holding them inside the empty bucket.
2. "I'll make the same sound in the water in this bucket. Listen the same way here. Listen for clicks. Are they louder or softer in the water? Put the golf tube in the water, then put your ear at the other end to listen. Is the sound louder coming through the air or through water?" Let the children make the scissor clicks for each other.

Note: If children in your class have gone fishing, they will understand now why people try to be very quiet when they are fishing.

3. *Do solid things carry sound?*

LEARNING OBJECTIVE: To become aware that sound vibrations can make solid objects vibrate to carry sound.

MATERIALS:

Table

Bottle caps, sticks, small rock, or anything that will make a noise on the table

Desirable: clock, aquarium pump motor, musical toy works

Finding out takes time. Keep the learning pace relaxed. Listen to the children's ideas.

SMALL GROUP ACTIVITY:

1. "We're going to make sounds for each other. Some may be very soft, so we'll have to listen closely." Tap one fingernail on the tabletop. "Does this sound seem loud? Now cover one ear with your hand and press the other ear on the table top while I do the same thing again."
2. "Which way was the sound louder, through the air or through the wooden table? Now put your ears to the table and close your eyes. Each one of you can take a turn tapping on the table with these things. We'll try to guess what made the sound."
3. Later compare sounds made by the clock, motor, or music works when held in the hand, placed on the table while heads are up, and on the table while ears are pressed to the table. Which sounds loudest?

4. *Will string carry sound?*

LEARNING OBJECTIVE: To become aware that sound vibrations can make taut string vibrate to carry sound.

MATERIALS:

Light string

Metal spoons

Bar of soap

Empty frozen juice cans, 2 for each telephone

1 ½" (4 cm) nails

Hammer

GETTING READY:

Cut string into 2 ½' (75 cm) lengths for spoon chimes.

Make a hole in center of each can bottom.

Cut string into 5' (1.5 m) lengths for metal can telephones. (String can be longer for use at home. Short lengths are best when many children will be using phones in the same room.)

SMALL GROUP ACTIVITY:

1. Dangle short string over a finger. "Do you think loose string will vibrate to make a sound if you pluck it? If it is held tightly by both hands? Try it both ways with your string." (Loose string vibrates too slowly for audible sound.)
2. "Let's add heavy vibrating metal to the string ends and listen to find out if tight string will carry sound." Tie a spoon to each end of string. Fold string in half; press folded end to ear. Lean over so that spoons dangle freely. Swing string to make spoons strike each other. Listen to the sounds traveling up the string.
3. Make metal can telephones with children. To stiffen string ends, pull them across soap bar. Put a string through the holes in two cans from the outside. Pull string into the can and tie around a nail. Wedge the nail inside the can. "Keep the string straight and tight. You talk into one can and your friend listens through the other can."

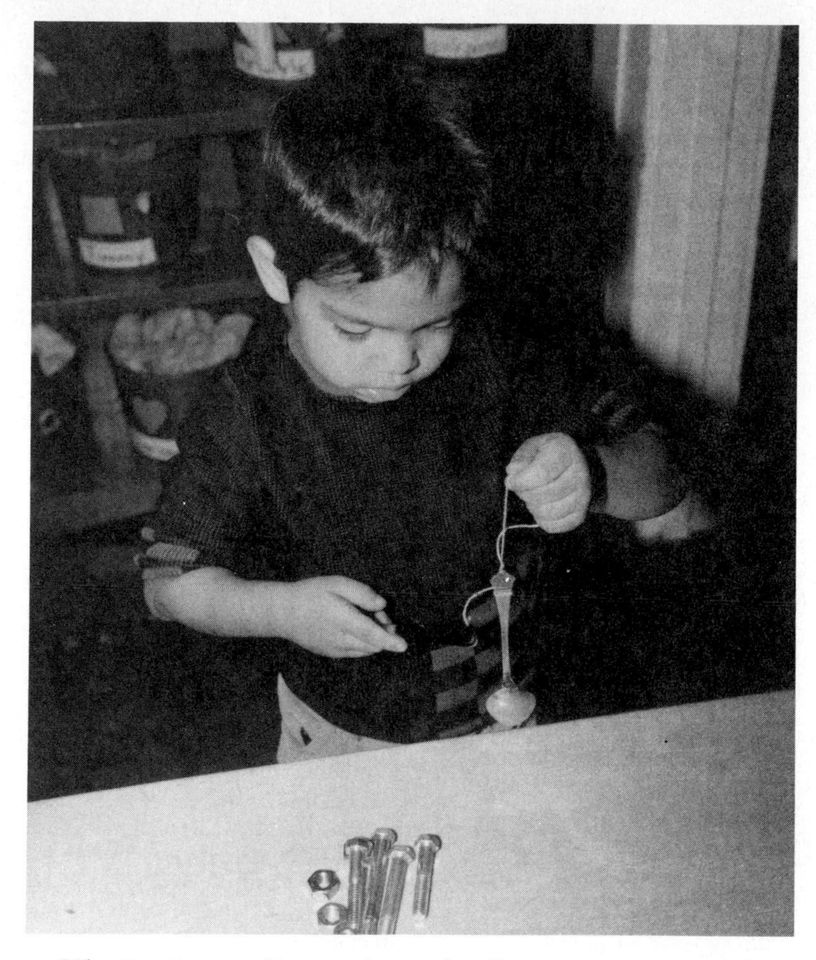

Vibration turns ordinary string and ordinary spoon into sound.

Group Discussion: Give each child a rubber band to pluck when it is limp and when it is stretched. What did the spoon chimes and metal-can telephone makers find out about how string vibrates to carry sound? How many things vibrated to carry the sound of their voices? (Air in cans, the cans, and the string.)

Read *Timmy and the Tin Can Telephone,* by Branley and Vaughn.

CONCEPT: Different sizes of vibrating objects make different sounds

1. Do different vibrating string lengths sound alike?

LEARNING OBJECTIVE: To hear pitch differences when different string lengths vibrate.

MATERIALS:

Three sizes of plastic boxes, such as 3", 4", 5" (8 cm, 10 cm, 13 cm)

Three sizes of small rubber bands

Empty covered boxes like a 2 pound (908 g) cheese box or small shoe boxes

Rubber bands large enough to stretch around the length of covered boxes

Small blocks

Autoharp or guitar (desirable)

GETTING READY:

Stretch small rubber band over each opened plastic box, the smallest band on smallest box, etc. (The effect won't be the same if same size rubber band is used on all three boxes. Try to find out why.)

SMALL GROUP ACTIVITY:

1. Pluck rubber bands on the plastic boxes. "Do these all sound alike? Which one sounds highest? Lowest? Try it."
2. "Now try this." Stretch long rubber bands around lengths of covered boxes. Slip block under band near left end of the box. Pluck the length of rubber band to the right of the block. Listen. Continue plucking and slide the block slowly toward the other end of the box. "Is pitch changing? Is it higher or lower? Does the whole rubber band length vibrate, or just the side being plucked? Is the part that vibrates getting shorter and higher?" Let children experiment, sliding the block and plucking. Can you play the scale this way? Play a tune?

Group Discussion: The autoharp shows the relationship between string length and pitch very clearly. (A real harp, and upright piano with the front panel removed, or a grand piano with the top raised will show the same thing.) If a guitar is used for this purpose, be sure to make clear which part of the string is vibrating, so that the relationship between vibrating string length and pitch can be understood.

2. Do different vibrating air column lengths sound alike?

LEARNING OBJECTIVE: To hear pitch differences when different air column lengths vibrate.

MATERIALS:

Carton of 8 empty 16 oz. (480 ml) bottles

Pitcher of water

Funnel

Masking tape

Sponge

Bicycle tire pump or bellows step pump

SMALL GROUP ACTIVITY:

1. "What is inside these bottles? What will happen if the air inside the bottles vibrates? When air moves across the top of the bottle it makes the air inside vibrate. Listen." Have a child use the pump while you direct a stream of air from the hose *straight* across the bottle top.
2. "We put some water in these bottles. Which bottle has space for more air? Let's see what happens when we make different amounts of air space vibrate."

It takes cooperative effort to make water organ music.

GETTING READY:

Experiment at home with levels of water needed to produce tones of the octave. Internal shapes of bottles vary, but these water levels may be about right:

(1) empty; (2) $1\frac{3}{4}$'' (4.5 cm); (3) $2\frac{1}{2}$'' (6.5 cm); (4) 4'' (10 cm); (5) $4\frac{1}{2}$'' (11.5 cm); (6) $5\frac{1}{4}$'' (13 cm); (7) $5\frac{1}{2}$'' (14 cm); (8) 6'' (15 cm).

Place band of tape around each bottle, with upper edge of tape at measurement level.

Let children pour water to marked levels of bottles, if possible.

3. Arrange bottles in order, 1 through 8. While a child pumps air, direct air from the hose across tops of bottles to play the scale. Try to play "Yankee Doodle." Let children play your bottle organ. (Put bottles in two cartons to prevent breaking, or keep bottles on the floor.)

Note: Listen to a beautiful recording of Romanian pan flute music. Zamfir plays this ancient pan-pipes instrument. The recording cover illustrates the graduated-size pipes he blows across.

PRIMARY INTEGRATING ACTIVITIES

Music

Many of the familiar children's songs are easy extensions of the learning about vibrations and pitch as they describe bell sounds and animal sounds. The concepts can be applied when rhythm instruments are used. "Charles, try holding your triangle by the string loop instead of with your hand. See if it vibrates more now." Prolong the life of school drums by suggesting that they vibrate better when lightly tapped than when thumped very hard.

Children can really make pleasant music with a classroom-made box harp. Excellent directions for making this no-cost instrument are given in *Music and Instruments for Children to Make,* by Hawkinson and Faulhaber (see Resources). Young children can help tape the box shut, but most of the remaining steps require the strength of an adult.

Directions for making no-cost drums, maracas, and castanets, as well as a low-cost xylophone are given in *The Science of Music,* by Melvin Berger (see Resources).

The following song is about the cause of sounds:

Sounds

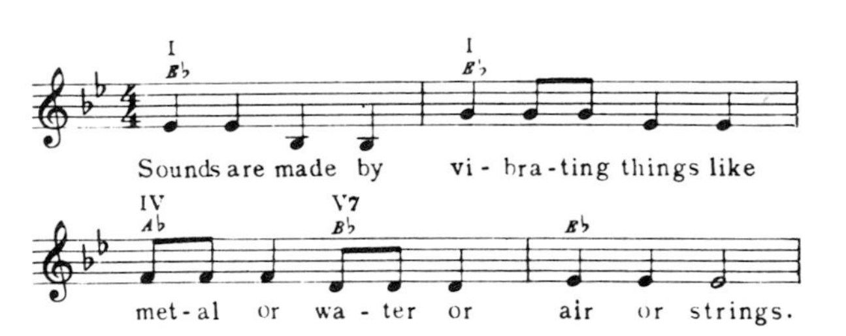

Math Experiences

Ordering By Pitch and Size. The mathematical relationship between the size and the pitch of vibrating objects is fun to explore. Sets of bells and sets of detached metal chimes, available commercially, can be placed on a molded framework to form a xylophone. The bells and chimes can be arranged by children in one-step tone order visually according to size, largest to smallest; or according to tone, lowest to highest.

Older children can have fun converting the bottle organ (page 240) into bottle chimes when they tap the bottles lightly with a pencil. They discover that the tone order is reversed. The bottle with little air space and a large amount of water produces a high tone when the air vibrates, and a low tone when the water vibrates. Because young children may be confused by this reversal, it would be wiser not to present this activity to them in two different ways.

Ideas for playing tunes on the water chimes are given in *Wake Up and Sing*, by Beatrice Landeck (see Appendix 1). However, Landeck's suggestion of using drinking glasses seems less safe for young children than using heavy bottles.

Stories and Resources for Children

ARDLEY, NEIL. *Sound and Music.* New York: Franklin Watts, 1984. A good resource for additional sound experiments, with scientific explanations of the results.*

BRANDT, KEITH. *Sound.* Mahwah, NJ: Troll, 1985. This reference can be directly read aloud to primary-grade children. It can be used selectively as a resource for younger children. Paperback.*

BRANLEY, FRANKLYN, & VAUGHN, ELEANOR. *Timmy and the Tin Can Telephone.* New York: Thomas Y. Crowell, 1959. Two children devise an early morning communication system between their houses that won't disturb their families.

______. *High Sounds, Low Sounds.* New York: Thomas Y. Crowell, 1967.

HOBAN, RUSSELL. *A Baby Sister for Frances.* New York: Harper & Row, 1964. Frances competes for parental attention by making and using a rattle from a gravel-filled coffee can.

KEATS, EZRA JACK. *Whistle for Willie.* Seafarer ed. New York: Viking Press, 1969. Peter has the thrill of discovering how to whistle.

______. *Goggles.* New York: Collier-Macmillan, 1971. Peter sends his voice through an empty drainpipe to confuse the big boys who are trying to catch him.

______. *Apt. 3.* New York: Macmillan, 1986. Varied sounds guide Sam to a sightless new friend, who knows his neighbors by the sounds he associates with them. Paperback.

KNIGHT, DAVID. *All about Sound.* Mahwah, NJ: Troll, 1983. Simple experiments and interesting capsules of information make this a useful reference for a broad range of ages. Paperback.*

POLLACO, PATRICIA. *Thunder Cake.* New York: Philomel, 1990. A young girl is distracted from her fear of thunder in this appealing story.

SCARRY, RICHARD. *Great Big Air Book.* New York: Random House, 1972. The story "Orchestra Practice" deals with sound waves traveling through the air. Amusing and informative.

WHITE, LAURENCE B., *Investigating Science with Rubber Bands.* Reading, MA: Addison-Wesley, 1969. Describes such things as making a rubber band banjo and changing pitch and volume.

______. *Science Toys and Tricks.* New York: Harper & Row, 1989. Some simple sound experiments are presented in this collection of activities. One lets you listen to a fly walk (but first you have to catch a fly!).

Poems (Resources in Appendix 1)

Many poems for children recreate sounds and extend the learning activities about sound. From *Poems To Grow On,* by Jean McKee Thompson, read "Kitchen Tunes," by Ida Pardue, and "The Storm," by Dorothy Aldis. From *Poems Children Will Sit Still For* read "Wind Song," by Lilian Moore.

Fingerplays

This fingerplay relates size to pitch and volume:

*Starred references, written at the young child's level of understanding, can help teachers with minimal backgrounds in science expand their knowledge base.

The Clocks

(slowly, loudly)	
When I'm a big tall clock,	(stand & stretch tall)
I go TICK TOCK, TICK TOCK, TICK.	(swing arms like a pendulum)
(faster, softer)	
When I'm a middle-size clock	(stoop over)
I go tick tock, tick tock, tick.	(swing arms faster)
(even faster, softer)	
When I'm a very small clock	(crouch low)
I go tick-tick-tick-tick-tick	(move hands fast)
So fast that I fall over!	

—Author unknown

Recall the metal-can telephone fun with this fingerplay.

Tin-Can Telephone

I called my friend on a tin-can phone.	(make cylinders with hands)
Two tin cans were joined by a string.	(extend little fingers to touch each other)
I put it to my mouth,	(put one cylinder up to mouth)
I put it to my ear,	(put other cylinder to ear)
I could hear my friend	
Say everything!	

Art Activities

Tambourines. Let children make crayon designs on two paper plates. Punch holes at regular intervals around the edges of the pair of plates. Children can lace the plates together partway, fill the tambourine with bottle caps, then complete the lacing. Plastic margarine tubs can also be used, but they are difficult to decorate.

Drums. Use salt or oatmeal boxes, or cans with air-tight plastic lids. Glue construction paper to the side of the drum, slipping rubber bands around the paper to hold it firmly while the glue dries. Children can decorate with crayons.

Horns. Let children glue and wind lengths of yarn around paper towel tubes. (The pressure children use when crayoning will squash most tubes.) An adult can punch an air-vent hole near one end. Help children fasten a single piece of waxed paper over one end of the tube with a tight rubber band. A reedy sound is made by tooting tunes into the open end of the tube.

Jingle Sticks. For each stick make a hole through the center of six bottle caps using a hammer and thick nail. Let children sand 8" (20 cm) lengths of dowel, 1 1/4" (4 cm) in diameter. Secure the dowel in a clamp or vise. Let children loosely attach three pairs of bottle caps (open edges together) near one end of the dowel with roofing nails.

Dramatic Play

1. A pair of tin-can telephones can add interest to the block corner airplane, ship, or spaceship play. They can also be fun to use in a pretend office, home, or hospital.
2. Medical play can include the experience of actually hearing each other's heartbeats. Use a real or improvised stethoscope to "gather" the sound so it can be heard more clearly. Insert the tube end of a soft plastic funnel into a short length of narrow-gauge plastic tubing. Place the funnel on a child's chest. Hold the tubing end close to *(not in)* the ear to hear the beats.

Thinking Games: Logical Thinking

Match My Sound. Gather pairs of matching items that will produce sounds, such as metal spring "crickets", two squares of sandpaper to rub together, small brass bells, jingle bells, two bottle caps to tap together, blunt scissors to click, small pieces of corrugated paper to scrape with a fingertip, pocket combs and waxed paper to blow on, two finger cymbals. Put one of each item in a teacher's bag. Distribute the matching soundmakers as evenly as possible into lunch-size paper bags. Fold down the tops and present a bag of secret sounds to each child at the table. While children keep their eyes tightly shut, the teacher reaches into her bag and makes a sound with one of the items. The children then search through their bags to see who can find things that produce a matching sound. Allow plenty of time for this game. The children may want to trade bags and begin again when you finish.

Sound Matching Tray. Use a divided cutlery tray and color film cans or small opaque plastic bottles to adapt a Montessori sensory training exercise. Loosely fill two matching containers with materials that make distinctive sounds when shaken, such as gravel, sand, pennies, rice, or puffed cereal. Mark matching pairs with matching numerals or letters on the container bottom as a self-correcting aid. Put one set of containers on one side of the tray; have children find the matching sound containers and place them next to their mates on the other side of the tray.

Field Trips

Look around your school area for a place that will produce an echo. An empty gymnasium or a high, windowless wall that borders an empty lot might be good places. Tell children that when they shout their names toward the walls the vibrations touch the walls and bounce back to them. Try it.

As you walk together for any special purpose, suggest walking very quietly so that you can listen for sounds to identify. An old church or auditorium with visible organ pipes, within walking distance of the school, is a good destination for this study.

Creative Movement

Children can show their perceptions of high or low sounds by moving to the tones of an instrument played by an adult. When the notes are low the children move in a low position close to the floor; when they are high the children move in a stretched-up position. They can also rise from a crouch to tiptoes in response to a slowly ascending scale, and return to the crouch as the descending scale is played.

SECONDARY INTEGRATION

Keeping Concepts Alive

Occasionally two of the loudest and most frightening sounds occur while children are in school: thunder and sonic booms. When lightning moves through the air, that air suddenly gets warm and it swells so fast that it vibrates. Thunder is the way we hear the vibration from that fast-swelling air far up in the clouds.

The unexpectedness of sonic booms adds to the scariness of that particular sound. Adults can agree with frightened children that surprise noises startle us. That acknowledgment and the example of our calm response are helpful to children. You could explain that when a jet goes very, very fast, it sometimes bumps into the air ahead that is already vibrating with its own sound. Sound travels fast, but sometimes a jet can travel faster than sound. People on the ground below hear that bump like a big explosion of air.

Relating New Concepts to Existing Concepts

Some of the suggested experiences relate to the children's existing concepts about the presence of air in empty-looking places and moving air pushing things (see chapter 7, What Is Air?). Friction is also involved in the production of some sounds, such as rubbing sandpaper together and unlubricated metal parts rubbing together to produce squeaks from the vibrations (see chapter 13, Simple Machines).

A nice way to relate sounds to animal life is to play the Tom Glazer recording *Now We Know (Songs to Learn By)* to hear "What Makes the Bee Buzz?"

Family Involvement

Any parent who is able to bring in and play a musical instrument can enrich the children's awareness of soundmaking. Families could be encouraged to lend things like a bell collection or wind chimes for the children to enjoy. Suggest events of soundmaking interest in the locality which they might visit with their children, such as a bell-ringing choir, a pipe-organ, or concerts.

EXPLORING AT HOME*

Touch and Watch To Tell. Join your child in making some discoveries about the sounds of home. He or she already knows that sounds are made by vibrating things. First do a touch test to find out if something is making a sound: Holding your ears shut with your fingers, touch a washer or dryer with your elbow when the appliance motor is on, then when it is turned off. Could you tell when it was making a sound, even with your ears shut? Let your child tell you about vibrating things' causing sound.

Now put a small jar of water on top of the appliance. Close your ears again. Watch the water in the jar when the motor is on, and when it is turned off. You can see the water vibrating in the glass, the effect of the motor's vibration and sound. Unstop your ears while you *hear* the sound of the motor and *see* the effect of the vibrating appliance.

Louder, Please!. Some things make sounds louder. Have fun with your child listening to yourselves whisper "Hello." Then close your ears with your fingers and whisper "hello" again. The sound is louder when you hear it inside your head. The vibrating air inside your mouth and nose, and the vibrating bony parts of your head amplify the sound. Enjoy listening to yourself!

RESOURCES

BERGER, MELVIN. *The Science of Music.* New York: Thomas Y. Crowell, 1988. A musician/science writer tells how sounds are heard by the ear, made by the voice and musical instruments, reproduced on records, tapes, and discs. Directions are given for making a drum, xylophone, and walnut shell castanets.

CARIN, ARTHUR, & SUND, ROBERT. *Teaching Science Through Discovery.* 5th edition. Columbus: Merrill, 1985.

HAWKINSON, JOHN, & FAULHABER, M., *Music and Instruments for Children to Make.* Chicago: Whitman, 1969. Excellent instructions for making a box harp that will withstand the wear and tear of child use. It will produce good clear tones.

*These suggestions may be duplicated and sent home to families.

15

Light

On the day of birth, infants are able to perceive light. Later the playpen explorer may try to catch a sunbeam. The toddler may try to pounce on a shadow. Growing young children are fascinated by the sparkle of reflected light and the beauty of the spectrum. The experiences that capture some children's closest attention, however, are those that help to allay their worries about darkness—the absence of light. The following concepts are suggested for investigation in this chapter:

- Nothing can be seen without light
- Light appears to travel in a straight line
- Shadows are made when light beams are blocked
- Night is Earth's shadow
- Everything we see reflects some light
- Light is a mixture of many colors
- Bending light beams make things look different

Children can experiment by examining a box full of darkness, using a flashlight beam to note the straight path of light, creating shadows, and looking through filters. Other experiences explain night and day, reflection, and refraction.

Introduction. To lead into these experiences with light, read one of these stories about the mastery of nighttime anxiety: *Bedtime for Frances*, by Russell Hoban, or *Switch on the Night*, by Ray Bradbury.

CONCEPT: Nothing can be seen without light

1. What can we see in a dark box?

LEARNING OBJECTIVE: To become aware that nothing can be seen without light.

A blanket-covered table adds intrigue to light experiments.

MATERIALS:

Pen flashlight

Extra batteries

Shoe box with cover

Small picture

Old heavy blanket

Low table

GETTING READY:

Cut a dime-size peephole in one end of the shoe box.

Cut a flap in top of box.

Paste picture to inside of box at end opposite the peephole.

Cover table with the blanket so it drapes to floor on all sides, making a dark place.

SMALL GROUP ACTIVITY:

1. "Let's stretch out on the floor and put just our heads into the dark place under the table."
2. "Here is a dark box. It has a hole in one end to look into. What do you see in there?" Pass the box to everyone.
3. "Now I am going to change something; then you can look into the box again." Turn on the flashlight and push it under the flap on the top of the box. "Can you see something in the box now? It was there before, but you couldn't see it. Let's try to find out why."
4. Remove the flashlight; pass the box again. "The picture is still in there. Can you see it? What happens when the light goes off? *We always need light to see anything. Nothing can be seen without light.*"

Be prepared to repeat this experience. Children are impressed with their ability to make the dark go away and come back at will. Listen to "Why Can't I See in the Dark?" sung by Tom Glazer on *Now We Know (Songs to Learn By).* (see Appendix 1.)

FIGURE 15–1

CONCEPT: Light appears to travel in a straight line

1. Does a flashlight beam curve around things?

LEARNING OBJECTIVE: To notice that light appears to travel in straight lines.

MATERIALS:

Pen flashlight

Extra batteries

3 sheets of white paper

Blanket and low table

Blocks

Masking tape

GETTING READY:

To prepare 3 screens: Stack the 2 blocks and tape the flashlight to the top block. Turn it on. Fold paper as shown in Figure 15–1. Stand it in front of the light. Cut a small hole where the beam hits the center of the paper. Cut holes in the same location in the other screens.

Line up screens a few inches apart so that the light beam can pass through all the holes. Cover table with blanket.

SMALL GROUP ACTIVITY:

1. "Let's find out how a beam of light travels. Lie down along the sides of the table. Put your heads under the blanket. I'll be at this end; no one will be at the other end."
2. Turn on the flashlight. "Notice the light shining on the blanket at the end of the table? It passes through each hole in the screens."
3. "Tom, will you put your hand in front of one hole, please? What happened? Did Tom's hand block the light or did it curve around the paper to shine through the next holes?"
4. "Take turns blocking the light beam as Tom did. Does the light curve around things or does light travel in a straight path?" Let children use the flashlight to check and recheck the path of light.

CONCEPT: Shadows are made when light beams are blocked

1. Does light shine through some things and not others?

LEARNING OBJECTIVE: To perceive light differently as it shines through different materials and to become aware that shadows result from blocked light.

MATERIALS:

Small flashlight

Blanket and low table

Waxed paper

Cardboard

Large shoe box

Clear glass jar

Water

Colored tissue paper

Clear vinyl folder

GETTING READY:

Cut a large opening in shoe box bottom.

Cut simple 3″ (7.5 cm) cardboard doll figure.

Fill jar with water.

Cover table with blanket.

SMALL GROUP ACTIVITY:

1. "Let's find out what light will shine through. You stretch out along that side of the table and I'll be across from you."
2. Put the box on its side with the opening facing the children. Shine the light through the opening. "Let's see if light shines through air."
3. Put the water jar in the opening. "Let's see if light shines through water."
4. "Let's see if light will shine through waxed paper in the same way." Cover the opening with waxed paper. "Does the light shine in one spot or spread out?" Try colored tissue paper and clear vinyl at the opening.
5. "Let's see if light will shine through this cardboard doll." Remove box; place doll so that it makes a shadow on the blanket.
6. Tell children to notice that light shines on only one side of the doll. "There is a dark place where the doll blocked the light. It's a shadow."

Group Activity: Move outdoors on a sunny day. Trace the outline of children's shadows with chalk on a paved area. As you trace, ask other children, "Is sunlight shining right through Amy or does she block the light? I'm tracing the darkness that Amy's body makes where the sunlight can't shine through her."

CONCEPT: Night is Earth's shadow

1. Why do we have day and night?

LEARNING OBJECTIVE: To begin to understand night's darkness as the shadow of the turning Earth.

Group Experience: (You will need a globe or ball, projector or flashlight, chalk, and a darkened room.) "We had fun blocking the sunlight to make shadows. Did you know that the whole Earth does the same thing? That is the way day and night happen. Let's pretend that this light is the shining sun. This globe is a model of our huge Earth. I'll put a chalk mark here on the globe to show where our part of the world is. Watch the mark as I turn the globe the way the Earth always slowly turns. Is the mark in the light now? Does the light curve around the globe to keep shining on our part of the world? No, the light shines in a straight line. Our part of the world is in a shadow now—the shadow of the other side of the Earth."

Continue to turn the globe, so that children can see the marked part of the globe alternately in the light and in the shadow. "What do we call the time when our part of the world is in the Earth's shadow? When it is in the sunlight?" Let the children turn the globe.

When your class encounters mention of the sun's *rising* or *setting* in stories, poems or songs, point out that such was the old, imaginary way of thinking about night and day. WE know that it only LOOKS like the sun is traveling around the Earth. We have daylight when our part of the traveling Earth spins toward the sun. After a while, rely on the class to correct such misconceptions they may find.

Listen to "Where Does the Sun Go at Night?" sung by Tom Glazer on *Now We Know (Songs To Learn By)*. (see Appendix 1.)

CONCEPT: Everything we see reflects some light

Introduction: Put a lighted purse-size flashlight into your cupped hands. Open your hands to let the children see where the light is coming from. "Is the flashlight making this light?" Hold a pocket mirror in your hands. "Is the mirror making a light?" Now shine the flashlight beam onto the mirror. Tilt the mirror so that the light is reflected on the walls or ceiling. "Which thing made the light that is bouncing up there, the flashlight or the mirror?" Turn the flashlight off and on several times so children can verify that the spot of light is *reflected* by the mirror to the ceiling only when the flashlight makes the light.

1. Do some things reflect light better than others?

LEARNING OBJECTIVE: To become aware that various materials reflect different amounts of light.

MATERIALS:

Strong sunlight, or spotlight study lamp

Large sheet of stiff white paper or cardboard

Aluminum foil pans

Shiny baking sheets

Colored construction paper: yellow, red, black

Dark carpet sample, velveteen, or bath towel

SMALL GROUP ACTIVITY:

1. Prop white paper on a chair seat at a right angle to the light source.
2. "Can you stand facing the light, holding a pan so it reflects light to the sheet of paper on the chair? See the reflection glow? Try to reflect light on the paper with these other things. See which ones reflect light well and which ones do not." Start with things that reflect well. Children may be discouraged as they try things if they don't get results right away.

Add more things to try, if possible. Small pocket mirrors are fun, but they may be hazardous. Tape edges of unframed mirrors if you decide to use them.

Read *The Rain Puddle*, by Adelaide Holl. What confused the animals?

CONCEPT: Light is a mixture of many colors

1. Can some things bend light to show its colors?

LEARNING OBJECTIVE: To observe the effect of light passing through a prism (or prism-like materials) and spreading into spectrum colors.

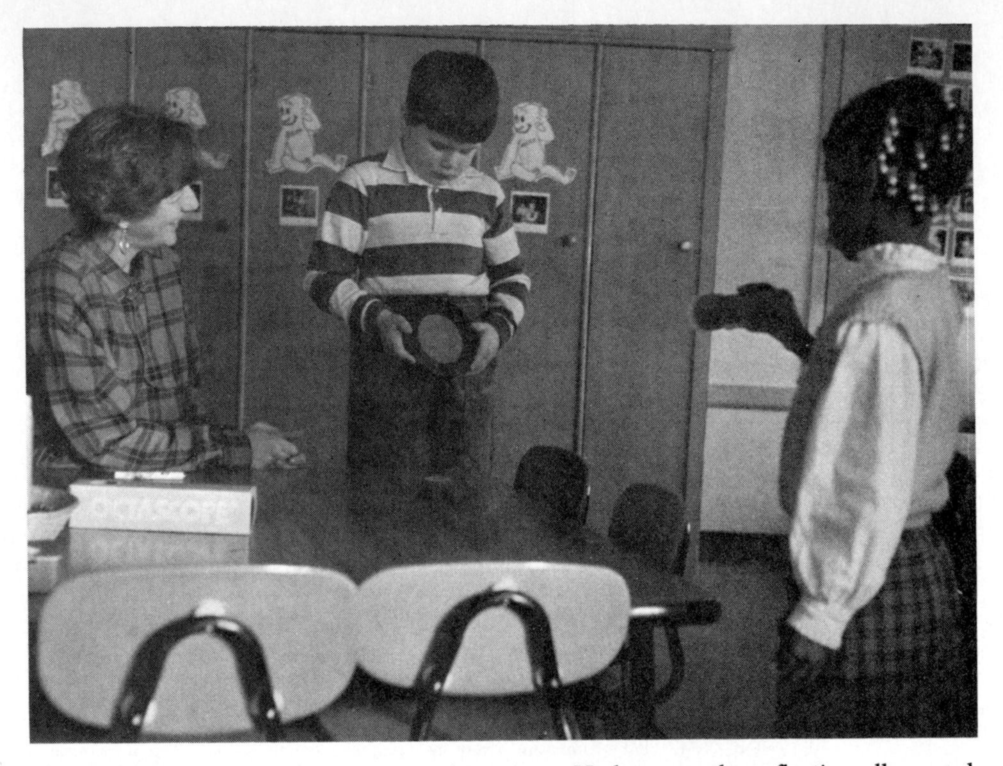

Billy catches Rose's flashlight beam in his mirror. He bounces the reflection all around the room.

MATERIALS:

Magazine

Strong sunlight or a projector

Small piece of thick plate glass or plastic, edges taped*

Prisms, or chandelier drops

Small, clear plastic boxes

Water

Soap bubble materials

Aquarium (desirable)

Diffraction grating cards, if available.

SMALL GROUP ACTIVITY:

1. "There is a secret in light beams. This magazine can help us find it." Hold the magazine; cut edges toward children, bound edge toward you. (Figure 15–2) "When pages are pressed together, does the edge look like one thick line? I'll bend it back toward me. Notice the line spread out to show many edges of paper. Try it."
2. Hold plate glass in front of light source. "When light comes through this glass it looks clear. Let's see how it looks when it is spread by glass that bends it."
3. Hold prism edge in front of light beam and rotate until a spectrum can be seen in the room. "Now you see the secret! Light really has seven colors in it (spectrum), but we only see them when light shines through something clear that is bent or curved." Let the children try it.
4. If possible, place an aquarium near sunny windows. Fill with water. Help children stand where they can see the spectrum through the aquarium corners.

*Do *not* use a magnifying glass.

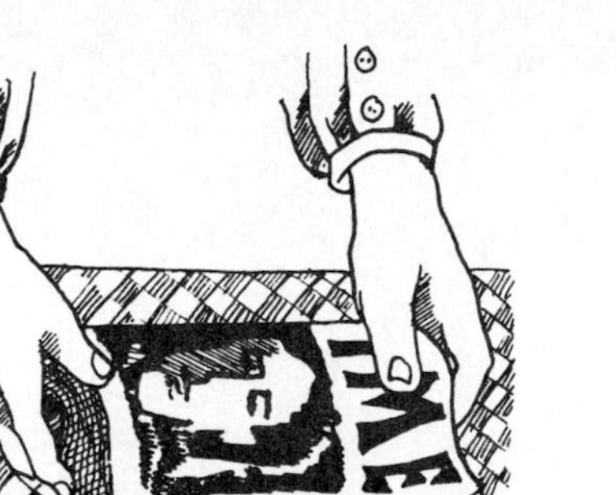

FIGURE 15–2

5. Let children experiment with water in small plastic boxes outdoors on a sunny day. Can they see the spectrum through the corners?
6. Try to make soap bubbles near a sunny window to see colors.
7. Let children examine diffraction grating cards to see the rainbow effects reflected by the patterns of lines.

Group Discussion: Talk about how raindrops sometimes act like prisms to form a rainbow in the sky. Listen to "Seven Colors Make a Rainbow" sung by Tom Glazer on *Now We Know (Songs To Learn By).*

CONCEPT: Bending light beams make things look different

1. How do things look through a curved drop of water?

LEARNING OBJECTIVE: To observe that things change in appearance when seen through transparent, curved materials.

MATERIALS:

Small bottles

Water

Medicine droppers

Waxed paper

Newspapers

Magnifying glasses or plastic "Card Lens" magnifiers

Sponges or paper towels

Small objects to examine

SMALL GROUP ACTIVITY:

1. "What happens when you put *just one* drop of water on your piece of waxed paper? Look at the newspaper through the drop of water. What do you see? Is the water drop curved or flat?"
2. "Now add more water to the drop. Is it still curved, or is it a flat spot of water? Do words beneath still look bigger? Why not? What has changed?"
3. "Absorb the water with a sponge and try again."

GETTING READY:

Fill bottles with water for each child.

Cut waxed paper and printed sections of newspaper into 3″ (7.5 cm) squares. Give each child a square of newspaper with waxed paper covering it.

4. Show children the curved profile of a magnifying lens. "Hold up the curved water bottle. Look at your finger through the bottle and water. How does it look?" (Light bends when it goes through curved glass or a curved drop of water. Things seen through them look different.) "We looked through curves that made things look bigger." Look at the curved grooves in the plastic "Card Lens" magnifiers. Make the lenses and magnifying glasses freely available for children's own independent explorations from now on.

Group Discussion: Try to find a simple microscope to show the children. "What could this mirror be used for? It reflects light up into the microscope." Read Millicent Selsam's book, *Greg's Microscope.* Try to make salt crystals to examine under the microscope the way Greg did. If a child in your class wears glasses, ask him how the glasses help him.

PRIMARY INTEGRATING ACTIVITIES

Music (Resources in Appendix 1)

Mention that laser light beams are used instead of needles to play compact disc recordings. Sing this song after children have used prisms (to the tune of "Good Morning Merry Sunshine"):

Prism Song

Good morning merry sunshine,
Your light comes straight to me.
When glass or water bend your line,
Your colors I can see.

Stories and Resources for Children

BRADBURY, RAY. *Switch on the Night.* New York: Pantheon Books, 1955. A child learns to control his fear of darkness.

BRANLEY, FRANKLYN. *Light and Darkness.* New York: Thomas Y. Crowell, 1975. The difference between things that create light and those that reflect light is clearly shown. Reassuringly emphasizes the constancy of the sun's light.

______. *What Makes Day and Night.* New York: Harper & Row, 1986. Earth's rotation is viewed from the vantage point of a spaceship, and a simple experiment at home personalizes what makes night and day. Paperback.*

______. *Eclipse: Darkness in Daytime.* New York: Harper & Row, 1988. Clear information for kindergarten/primary-grade level.

BROEKEL, RAY. *Experiments with Light.* Chicago: Children's Press, 1986. Includes enrichment level experiences with refraction and color, as well as basic information about light.*

*Starred references, written at the young child's level of understanding, can help teachers with minimal backgrounds in science expand their knowledge base.

———. *Experiments with Water.* Chicago: Children's Press, 1988. Two refraction experiments described in this book link water and light topics.

CONAWAY, JUDITH. *More Science Secrets.* Mahwah, NJ: Troll, 1987. Directions for making a solar-powered pinwheel and 'solar mittens' are included in this activities collection. Paperback.

CREWS, DONALD. *Light.* New York: Greenwillow, 1981. Light and darkness in the city and country. A simple text for young preschoolers.

DOROS, ARTHUR. *Me and My Shadow.* New York: Scholastic, 1990. Simple, clear explanations are presented for night as Earth's shadow, and the phases of the moon as its own shadow. Shadow tag and other shadow fun are included.

GOOR, RON AND NANCY. *Shadows Here, There and Everywhere.* New York: Thomas Y. Crowell, 1981. Intriguing photographs show that shadows can be useful.

HOBAN, RUSSELL. *Bedtime for Frances.* New York: Harper & Row, 1960. Monstery shapes become familiar chairs and robes when father turns on the light in Frances' bedroom.

HOLL, ADELAIDE. *The Rain Puddle.* New York: Lothrop, Lee & Shepard, 1965. Evaporation of water and reflections confuse some farm animals.

LIONNI, LEO. *Little Blue and Little Yellow.* New York: Ivan Oblensky, 1959. A simple color story.

SCHNEIDER, HERMAN & NINA. *Science Fun with a Flashlight.* New York: McGraw-Hill, 1975. Shadows, the spectrum, reflected light, and the nature of a light beam are explained well.

SELSAM, MILLICENT. *Greg's Microscope.* New York: Harper & Row, 1963. Greg's new microscope reveals exciting information about everyday materials.

SIMON, SEYMOUR. *Mirror Magic.* New York: Lothrop, Lee & Shepard, 1980. Children experiment with reflected light.

———. *Shadow Magic.* New York: Lothrop, 1985. Children playing with shadows are led, through questioning, to discover basic information about shadows. Directions for making a shadow clock are included.

SONNEBORN, RUTH. *Someone Is Eating the Sun.* New York: Random House, 1974. The old turtle has experienced an eclipse before, so he can dispel the sun-eating theory of younger animals.

WYLER, ROSE. *Raindrops and Rainbows.* Englewood Cliffs, NJ: Simon & Shuster, 1989. Simple text and experiments explore aspects of separating light into rainbows.

Storytelling with Lights

1. Tell Leo Lionni's story, *Little Blue and Little Yellow,* using a small study lamp and blue and yellow plastic paddles to blend and separate the color families in the story. Let the children experiment with the light and paddles afterward. Have a sheet of white paper under the lamp. A set of colored plastic construction squares can also be used this way.
2. Use a small high-intensity lamp and three-dimensional objects to animate and tell a simple shadow story. Do this on the floor of a darkened room with the children sitting in a circle around the story setting. Small dolls can be children playing shadow tag outdoors. A cardboard tree or a wall of blocks can form a shady place for the dolls to rest after the game. A cardboard cloud could pass between the dolls and the lamp to perplex the story children. They can't find their shadows for the tag game and think they've lost them. Walk the dolls over to the children watching the story to ask them for help finding the shadows.

Poems (Resources in Appendix 1)

There are many poems about light.

From *A Light In The Attic,* by Shel Silverstein:
"Shadow Race" and "Reflection"
From *Poems to Grow On,* by Jean McKee Thompson:
About shadows: "Shadow Dance," by Ivy O. Eastwick
About shadows: "Kick a Little Stone," by Dorothy Aldis
About reflections: "Mirrors," by Mary McB. Green
About the spectrum: "I Wonder," by Virginia Gibbons
From *Poems Children Will Sit Still For,* compiled by Beatrice deRegniers:
About reflection: "A Coffeepot Face," by Aileen Fisher
About shadows: "8 A.M. Shadows," by Patricia Hubbell

"My Shadow," by Robert Louis Stevenson, from *A Child's Garden of Verses,* may need a bit of translating for children who know babysitters, but not nursemaids. It should be part of their poetry experience.

Art Activities

A Reflecting Collage. Let children make collages with pieces of foil gift wrap, aluminum foil, sequins, glitter, and gummed stars.

Reflector Hanging. Let the children cut pairs of simple shapes from foil paper or aluminum foil. Help them place three or four shapes face down in a line. Put a few drops of white glue in the center of each shape. Place a 15" (38 cm) length of string on the glue spots, then let the children cover each shape with its mate. Tie a gold plastic thread spool to the bottom of the string. Tape to the ceiling or to a door frame.

"Stained Glass" Medallions. Cut apart separate rings of plastic "six-pack" holders from frozen juice or drink cans. Pierce a hole at the top of each ring with a heavy threaded needle. Tie a thread loop. Let children dot glue around the ring and press colored tissue paper and cellophane over it; then trim away excess paper. Hang the medallions in the windows. Does light come through?

Waxed Paper Translucents. Let children cut shapes from colored tissue paper and arrange them as they wish between two sheets of waxed paper. Children seal the papers together by placing them on a newspaper-covered food warming tray and rubbing firmly across the waxed paper with a pizza roller.

Color Blending. After reading *Little Blue and Little Yellow,* by Leo Lionni, children may be interested in blending paint colors. Younger children can try two primary colors at a time. Older children can try to create all the colors of the rainbow with three primary colors.

Dramatic Play

Ship Play. The crew of a building-block ship will enjoy using a periscope. Let them see how mirrors reflect light to create this interesting effect. Instructions for making a periscope are found in *Sun and Light,* by Neil Ardsley (see Resources).

Shadow Fun. Draw the shades during active playtime on a rainy day and set up a projector at one end of the room. Turn on dancing music and let the children enjoy creating shadows as they dance in a designated area. Play shadow tag with the children outdoors on a sunny day.

Table Play

Reflection/Refraction Attraction. Follow up the reflection and refraction experiences by making as many of these fascinating toys and practical items available as you and your class can bring in to share. Kaleidoscopes, octascopes, and homemade milk-carton periscopes, all depend upon mirrors within to create intriguing patterns and views from scraps. The Dragonfly Optiscope has 25 planes in its lens to imitate that insect's view of the world. Diffraction-grating lenses and novelty items produce rainbow effects. One popular miniature figure has such a lens that creates an appearing/disappearing face. These items are usually sold in nature stores and museum shops. Add old binoculars, varied plastic card magnifying lenses and other magnifying lenses, and a container of small things to enlarge and examine.

Creative Movement

Be My Mirror. Suggest to the children that they pretend to be your reflection in a mirror. They silently move as you move. You may be surprised at the number of movements you and the children can invent while you are seated, using your head, neck, shoulders, arms, hands, and fingers. This is a good activity to calm a restless group.

Thinking Games: Imagining

What If. "What if none of the lights could be turned on in your house one night?" How could things be identified in the dark? Pass a feeling bag containing several objects that can be identified with closed eyes by feeling or smelling.

SECONDARY INTEGRATION

Keeping Concepts Alive

1. Point out reflections when you notice them in the classroom: in doorknobs, in a child's shiny brass button, in the spoons at lunchtime, or in the rain puddles on the playground.

A quiet moment alone with the new microscope fascinates Rose.

2. Hang a prism in a classroom window that gets direct sunlight. Experiment with balance and location to produce a good spectrum.
3. Use the opportunities that arise to comment on the need for light to see well: when raising the shades after rest time, when looking closely at a scraped knee, and so on. Comment that we need light to see our wonderful world.
4. Store magnifying glasses in a place where children will have access to them when they want to examine something. If you have a classroom microscope, let the children know that you will get it out for them when they want to examine a specimen. Talk about reflecting more light onto the specimen with the adjustable mirror.
5. Keep track of the length of the shadow made by a special tree or building near the school playground. Use a chalk mark on a paved surface, or stones on a grassy area, to compare its length when you are outdoors in the morning, at noon, and in the afternoon.

6. When the subject of lasers is brought up by children in terms of the destructive weapons they see in television programs, or in certain toy advertisements, balance these ideas with the constructive uses of laser light. Tell them that this form of synthetic light is used by doctors to heal people. It is a powerful tool with many good uses. It is also used to "read" the special marks on grocery packages at the supermarket checkout counter they often pass. A tiny laser beam shines on compact discs to produce sound.

Relating New Concepts to Existing Concepts

Relate the bending of light beams to the surface tension experiences (see chapter 8, What Is Water?). Recall that the pull on the outside of the water drop makes it curved. Light coming through this curve of water bends and changes the way in which we see through it. Surface tension keeps soap bubbles curved, and lets us see rainbow colors of spreadout light beams. In caring for classroom plants, remind children that all plants need light to stay alive and grow. Animals and humans need light to live, too, because they need plants for food.

Family Involvement

Encourage families to take their children to these features, if they are available in your community: an observatory, a lighthouse, stained glass windows, electric-eye door openers. Point out solar panels on home roofs that store the sun's energy for heating or convert solar energy to electricity. Families may have pocket calculators that use photovoltaic material to change light into electricity. Families can show children the laser scanner units at the supermarket and point out how the laser signals the price of groceries to the computer cash register.

EXPLORING AT HOME*

Sharing Shadows. One of the best places for your child to learn about shadows is at home with you after dark. She or he already knows that shadows form when light can't shine through something. So find a strong flashlight or a spotlight-style lamp that *you* control, and join your child in a darkened room to try to share these explorations:

1. Shine the light on your child's back as he or she stands close to the light. Notice the shadow on the wall ahead. Notice how the shadow changes as your child slowly walks toward the wall.
2. Do the same thing, but this time with the light at floor level. Try it again with the light held high above the child's head. Enjoy noticing how each change makes a difference in the shadow size or location.

*These suggestions may be duplicated and sent home to families.

3. Add a second light source, shining from a different direction. How many shadows are there now? (Remember to point out shadows of players if you go to a ball game under the night lights.)
4. Hold two layers of *waxed paper* in front of the light. Notice how your child's shadow changes when some of the light is blocked by the cloudy paper. Mention that shadows change like this when clouds block the sunlight from us.
5. Have fun making shadows of familiar objects: toys, forks, combs and whatever else you think of. Make the shadow shapes you learned to make as a child.

When you are outdoors together on a sunny day, you might want to protect your face from too much sunlight by wearing a shadow maker: a visor or sun hat!

RESOURCES

ARDLEY, NEIL. *Sun and Light.* London: Franklyn Watts, 1983.
BAINS, RAE. *Light.* Mahwah, NJ: Troll, 1985. Includes a brief note on laser light.
BRANLEY, FRANKLYN. *The Sun: Our Nearest Star.* New York: Harper & Row, 1988. Includes illustrations of solar panels using the sun's energy to heat a house. Paperback.
LEVENSON, ELAINE, *Teaching Children About Science.* Englewood Cliffs, NJ: Prentice-Hall, 1985. Periscope-making, good experiments & explanations, chapter 7.
MACAULAY, DAVID. *The Way Things Work.* Boston: Houghton Mifflin, 1988.
MIMS, FORREST. *Lasers: The Incredible Light Machines.* New York: David McKay, 1977.

Teaching Resource

SUNPRINT KIT. Create images using sunlight and light-sensitive paper. Museum stores or nature stores carry these. Also may be ordered from Lawrence Hall of Science (see Appendix 4).

16

Electricity

Young children often encounter electric charges during dry times of the year. A sneakered child scuffles across a nylon rug and receives a small shock upon touching the doorknob. Sweaters crackle as they are pulled on and hair flies into a halo when it is brushed. Why? The answer can help conquer fears about that giant electrical charge—lightning. The following concepts underlie the experiences in this chapter:

- Friction can build electric charges
- Charged electrons make sparks when they jump
- Current electricity travels in a loop
- Current electricity does not move through some things

An initial group activity is suggested to label the effects of electric charges. A demonstration and discussion introduce the idea that current electricity is pushed through wires by batteries or generators. It concludes with a safety discussion. Children can try to complete a simple circuit, add a knife switch to the circuit, and experiment to find materials that conduct electricity.

Note: The experiences with electric charges (Part 1) will not be successful in a humid atmosphere. It is best to save these experiences for dry weather.

PART 1: ELECTRIC CHARGES

Introduction

Show the children a 2-inch square of plastic wrap, held between your thumb and forefinger. "What do you think might happen when I let go of this plastic?" Find out.

"Let's see if friction might make a difference in what happens to the plastic." Rub the plastic on a piece of wool or nylon fabric. Pick it up again and let go. "Look at this. I have let go, but the plastic hasn't let go this time. Friction did change some things in the plastic, things so very tiny they are invisible. They are *electrons*. Electrons are part of everything—*even part of us*. Friction gets the en-

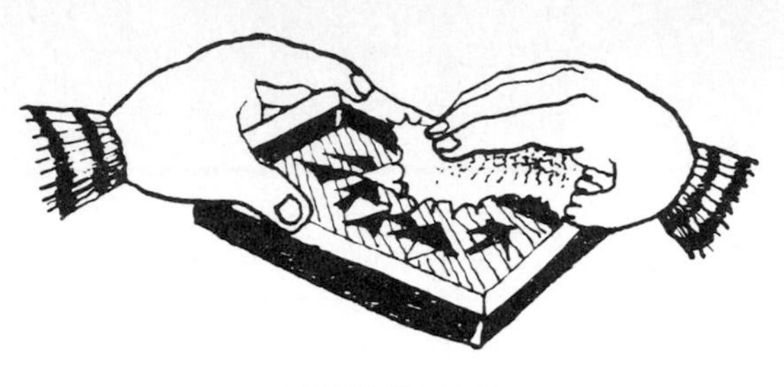

FIGURE 16–1

ergy of the electron ready. That is called *charging,* or building an electric charge. Often it is also called *static electricity.*" (The children may already have many ideas about static electricity.)

Give children their own pieces of plastic to try. A sample of nylon carpeting or garments that children are wearing can be good synthetic fibers to rub the plastic against.

"Electrons are part of everything, but we don't notice them until they are charged. You can have fun using friction to build an electric charge and make some little things jump."

CONCEPT: Friction can build electric charges

1. Can we build up an electric charge to make things move?

LEARNING OBJECTIVE: To observe the effects of electric charges on noncharged materials.

MATERIALS:

Clear, flat plastic boxes

Tissue paper

Combs (plastic or nylon)

Small pieces of cotton thread

Scraps of wool, fur, silk, or nylon

SMALL GROUP ACTIVITY:

1. Ask children to tear up tissue paper into rice-size bits. Put some in the boxes.
2. "Let's rub a box top with a bit of fabric very fast (Figure 16–1). The friction can make static electricity move. It's called building a charge. Watch the bits of paper. They move when charged electrons move."
3. Suggest turning the boxes over after tissue bits jump and cling to the top. Try to build a charge on the bottom.
4. "Now try to build a charge on the comb with the fabric. Hold it near the thread. What happens? Did the thread jump when the static charge jumped?"
5. "Hold the charged comb near your hair. Does some hair move with the moving electricity?"

2. Can charged electrons make things stick together or push away?

LEARNING OBJECTIVE: To observe the clinging effect of unlike-charged materials; the repelling effect of like-charged materials.

MATERIALS:

Foam meat trays

Small, long balloons

Wool, silk, or fur

String

GETTING READY:

Mark children's initials on balloons with felt-tip pens.

Tie a foot of string to each of two balloons.

SMALL GROUP ACTIVITY:

1. "Will this tray stick to a wall by itself? Find out."
2. "Let's see if static electricity will help." Rub fabric very fast on the tray's underside to build up a static charge. "Now will it stick to the wall?"
3. Carefully build a charge on balloons. Will they stick to the walls?
4. Now build a charge on string-tied balloons. Holding the string, bring charged sides of balloons together. (Bits of fiber stick to charged sides of balloons.) Do they cling together or push each other apart?

Did this remind someone of the way like ends of magnets push away and unlike ends cling? These are negative and positive electrical charges. Like charges push apart and unlike charges cling.

CONCEPT: Charged electrons make sparks when they jump

1. Can we make a tiny lightning flash?

LEARNING OBJECTIVE: To begin to understand the nature of lightning.

Introduction: "Let's see if we can make a tiny lightning flash in our dark place. We'll pretend that balloons are storm clouds full of water. Water drops are blown about and rubbed together in a storm cloud to build up a big charge of electricity."

MATERIALS:

Small, long balloons

Small wool or nylon rug, or carpet sample

Heavy blanket, table

GETTING READY:

Drape table with the blanket to make a very dark place.

Place rug on the floor under the table.

Blow up balloons.

SMALL GROUP ACTIVITY:

1. Show children how to stroke the balloons on the rug to build the electric charge. Count 20 strokes with them.
2. "Now pretend my finger is another cloud or the Earth. I'll bring the charged balloon close to it. Watch! Listen! Did you see the spark jump to my finger? Now you try it."

Group Discussion: "You can make a tiny lightning flash at home in a room that is almost dark. Scuffle your feet on a wool or nylon rug, then touch something metal. You can feel, hear, and see a small charge of electricity jump from yourself to the metal."

Success for Marc and Michael. Their light switch works!

"A small spark like this is not dangerous, but lightning from storm clouds is very powerful. That is why we take shelter in a building or car during an electrical storm."

PART 2: CURRENT ELECTRICITY

Introduction

Use an air pump, a flashlight battery, and the simple circuit materials to introduce the topic of current electricity. Show the children the air pump. "Do you remember how we used the pump to push air into the bags (inner tube, air mattress)? Does the pump make the air? No, it pushes along the air that comes into it through this air hole."

Show the battery and dry cell. "These things also push along something we can't see." Some children will recognize them as batteries that make flashlights shine and electrical toys work. "Batteries don't *make* electricity. They push the electricity that is already in something to make it flow along. When it moves we call it electric current. It flows through wires in a special way. Small batteries like these can push only a small amount of electricity through the wires, so it is safe enough for children to touch. *The electricity that is pumped by giant generators into wires for houses and schools is so powerful that it is too dangerous for people to touch.* If the powerful electricity does not move through the wires the right way, it can burn things or hurt people very badly if they touch it." Children can find out how to make a safe amount of electricity flow through the wires to make a tiny light shine.

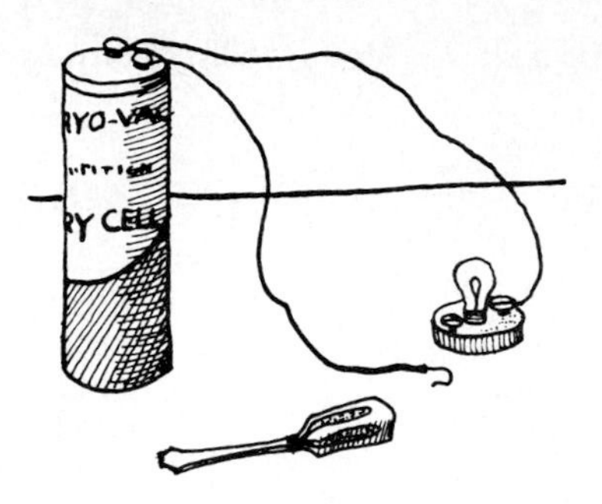

FIGURE 16–2

CONCEPT: Current electricity travels in a loop

1. Can we complete a simple circuit?

LEARNING OBJECTIVE: To find out (with safe materials) that electric current only moves in a complete loop.

"Current electricity can move only in a loop pathway. You can find out what this means by trying to light a small bulb."

MATERIALS:

For two children at a time:

#6 dry cell ($1\frac{1}{2}$ volts)

Insulated copper wire

$1\frac{1}{2}$ volt light bulb

$1\frac{1}{2}$ volt socket

2 tiny screwdrivers

GETTING READY:

Cut wire 12" (18 cm) long.

Cut and peel off $\frac{1}{2}$" (2 cm) of insulation from each end of wires.

Bend exposed wire into hooks by folding around a pencil tip.

SMALL GROUP ACTIVITY:

1. Have materials out for one circuit. Show the children how to loosen a dry cell terminal cap, hook the end of wire beneath it, and tighten the cap.
2. Loosen (don't remove) a socket screw; hook the other end of the wire around it and tighten it (Figure 16–2).
3. "The light didn't go on yet. We need to make a complete loop for electricity to move on. It has to flow from the dry cell, through the light socket and back to the dry cell. See if you can complete the loop to make the light go on."
4. Later ask if children can find ways to turn the light off.

Group Discussion: With a finger, trace the path of the completed circuit. Start at one terminal, follow the wire to the socket. Remove the bulb to show the metal pathway in the socket. Follow the second wire back to the dry cell. Have the children trace the path.

Explain that if a wire or piece of metal touches both terminals on top of the dry cell, the current will travel in just that small loop, and it would use up all the power too fast. The dry cell could not be used to make the lights shine after that. No metal or wire should ever touch both terminals. Try to keep the equipment available for children to return to many times.

2. *How does a switch stop moving electricity?*

LEARNING OBJECTIVE: To find ways (with safe materials) to complete and break an electric circuit.

MATERIALS:

#6 dry cell

Insulated copper wire

$1\frac{1}{2}$ volt light bulb

$1\frac{1}{2}$ volt socket

2 tiny screwdrivers

Knife switch

Flashlight

Wall switch and cover

GETTING READY:

Cut a 6″ (15 cm) piece of wire, trim away insulation, and bend ends into hooks.

SMALL GROUP ACTIVITY:

1. Let children complete a simple circuit.
2. "What happens when you push a light switch? Let's find out." Show how the knife switch opens to make a gap in its metal pathway and closes to complete it.
3. "Can you find a way to add the switch and a new wire to the completed circuit?"
4. "There is a complete loop for electricity to move through with the handle down. What happens to the light when the handle is up?"

Let the children open and examine the parts of the flashlight. The metal case takes the place of wires to make a pathway. Compare the wall switch parts with those of the knife switch. Could it work in the same way to complete and break the circuit?

Group Discussion: Show the children an extension cord. How many wires does it have side by side? Look at the two prongs on one end and the two holes on the other end. Why do you think two wires are needed? Would electricity flow if the extension cord were plugged into itself? What is missing? Children must never play with electrical outlets. Why?

CONCEPT: Current electricity does not move through some things

1. *Which materials let current electricity pass through; which do not?*

LEARNING OBJECTIVE: To discover (with safe materials) which materials conduct electricity.

MATERIALS:

#6 dry cell

$1\frac{1}{2}$ volt socket and bulb

Electrical wire

Roofing nails, or large head tacks

Small blocks of wood

SMALL GROUP ACTIVITY:

1. Let children set up a switch circuit, winding exposed ends of two wires around the two nails instead of connecting to a switch.
2. "Why isn't the light shining now?"
3. "Let's see if we can complete a circle pathway by putting a key across the two nails. How do you know if electricity moves through it? If it does, put the key on the *yes* tray."

A visiting volunteer helps the group add a switch to their circuit.

Test materials: string, rubber bands, sticks, long nails, keys, plastic spoon, 1" (2.5 cm) tile squares, stones, leather, glass, paper clips, paper

Glass or ceramic insulator

Trays or box lids

GETTING READY:

Hammer 2 nails an inch (2.5 cm) apart into blocks of wood.

Label the trays *yes* and *no.*

4. Let children try placing other materials across the space between the two nail heads to discover which things let electricity pass and which do not. Use the terms *conductor* and *insulator.*
5. Change one wire, using a new wire with the insulation still covering the end. Does electricity pass through, even though the circuit *looks* complete? Why not?

Group Discussion: "It is important to know about two other things that electricity can flow through—people and water. We must never touch a light switch when we have wet hands or when we are standing on a wet floor or in the bathtub. If a wire were loose inside the switch, powerful electricity would pass from the wire through the wet person. That is also why lifeguards tell swimmers to come out of the water when an electrical storm starts."

PRIMARY INTEGRATING ACTIVITIES

Music (Resources in Appendix 1)

Try singing about static electricity with the following song (to the tune of "I'm A Little Teapot").

The Static Song

When we scuff a rug with feet so quick,
We touch some metal and get a prick.
If we do the same thing in the dark,
When static jumps, we see a spark.

Stories and Resources for Children

BERGER, MELVIN. *Switch On, Switch Off*. New York: Thomas Y. Crowell, 1989. Clear, beginning ideas about how electricity is generated and then used in homes. Simple circuits and the light bulb are explained. Directions for creating a tiny electric current are included.*

BRADBURY, RAY. *Switch on the Night*. New York: Pantheon Books, 1955. Your library may help you find a copy of this simple story of light, electric and otherwise helping a child overcome fear of darkness.

BRANLEY, FRANKLYN. *Flash, Crash, Rumble, and Roll*. New York: Harper & Row, 1988. Describes static electricity charges building into lightning.

CHALLAND, HELEN. *Experiments With Electricity*. Chicago: Children's Press, 1986.

HOBAN, RUSSELL. *Arthur's New Power*. New York: Thomas Y. Crowell, 1978. Arthur's family has power problems that Arthur solves by building a water wheel to generate electricity.

MARKLE, SANDRA. *Power Up*. New York: Atheneum, 1989. The revealing and inexpensive experiments in this book are safe enough for older children to do alone. Only "D" batteries, folded aluminum foil "tape wires," and a few other simple materials are required to explore electric current.

PODENDORF, ILLA. *Energy*. Chicago: Children's Press, 1982. Illustrates many kinds of energy that can be used to generate electricity.

SCARRY, RICHARD. *What Do People Do All Day?* New York: Random House, 1968. Electricians are installing an electric circuit in a new house. Safety features can be seen. The continuous pathway of wires running behind walls is well illustrated.

SCHNEIDER, HERMAN AND NINA. *Science Fun with a Flashlight*. New York: McGraw-Hill, 1975. Contains a simple description of how electricity from a battery lights a bulb.

SMITH, KATHIE B., & CRENSON, VICTORIA. *Thinking*. Mahwah, NJ: Troll, 1988. Briefly describes how all the body's muscles and organs are "wired" to the brain.

WADE, HARLAN. *Electricity*. Milwaukee: Raintree Children's Books, 1979.

Poems (Resources in Appendix 1)

From *Poems To Grow On*, by Jean M. Thompson, read Dorothy Aldis's poem, "The Storm." Point out that lightning and thunder are part of the same event. We see the spark before we hear the vibration.

*Starred references, written at the young child's level of understanding, can help teachers with minimal backgrounds in science expand their knowledge base.

Jason just needs a little help to screw his wire to the switch.

Fingerplay

Adapt a familiar fingerplay to include the children's new awareness of the biggest static spark of all, lightning.

This little boy	
crawls into his bed.	(Hold up right forefinger)
He pulls the covers	(Place it on open left palm)
over his head.	(Fold left fingers over palm)
He turns off the light	
to sleep through the night.	(Move left thumb down)
Till a loud thunder crash	
and a bright lightning flash,	
Make him jump out of bed	
to watch static, instead.	(Release right forefinger)

Art Activities

Collage. Let children cut out and paste catalog pictures of electric appliances to make a "We Use Electricity" collage.

Wire Art. Heavy telephone cables consist of bundles of copper wires coated with insulation of varied colors. Telephone repairmen can give you discarded lengths of cable. Twelve-inch (30 cm) lengths of wire are easy for children to work with. Pull out wires from the bundle as needed. Discuss the nature of the material as you give it to the children. Let them use it for lacing or weaving through pieces of wire screen or burlap, or for making a sculpture. For sculpture bases, either staple a loop of wire to scrap lumber, or let the children poke their wires into chunks of pressed foam. (It is fairly easy to saw large pieces of molded foam packing material into usable sizes.) Show the children a few techniques, such as winding wire around a crayon to make a spiral, or threading other materials (paper or foam chips) onto the wires.

Dramatic Play

1. Provide a solar construction kit for independent activity. Available from Seventh Generation (see Appendix 4).
2. Let the children help you convert a cardboard carton into a tabletop "control panel" for their imaginative play. Use the knife switch and light on the outside of the box, attaching them to wires poked through the box to connect with the dry cell inside. Draw dials and gauges on the outside of the box; add a paper plate wheel.
3. Wire a shoe box dollhouse by cutting doors and windows in the sides of a shoe box. Put two wires through small holes in the box top; connect them to the dry cell and the light bulb socket to make a ceiling light. Children enjoy making furniture for the house from egg carton sections, and dolls from clothespins and pipecleaners.

Food Experience

A toaster provides a safe vantage point from which to observe and use electricity in a food experience for young children. The children can help to prepare buttered toast for a snack or for lunch while watching what happens when electricity passes through a special kind of wire. The glowing wires are held in place with insulators. Can the children see them? Check the safety of using the toaster in your room beforehand.

Thinking Games: Imagining

1. Play a "What if?" game to think about the many ways we depend on electric power. "What if we couldn't use any electricity tomorrow? How would we start the day? Would an electric alarm clock wake us up?" Follow through the day as the children know it. What can they do to use electricity conservatively?
2. Play "Bulb, Wires, and Battery." Sit with the class in a circle on the floor. Ask a child across from you to be the light bulb and sit with arms folded. You be the battery with clenched fists. All of the other children hold hands and become two wires. Let the "wire" children next to you put one hand in your fist. Demonstrate how a squeeze from your fists can be passed from one child to the next like electrons moving down the path of a circuit. When the charge reaches the "light bulb," the child raises his arms to light up. If the "wires" become disconnected at the "battery" or "bulb," the bulb does not light and the arms are folded again.

3. Add an electrical imagery experience to the "Bulb, Wires, and Battery" game. If feasible, have the children stretch out on the floor. If not, seated children can be guided through the experience as well. Quietly and *very slowly* say, "Now let yourselves feel comfortable and loose and easy. Let your eyelids slide down . . . so you can see inside yourselves in your imagination. Feel your lungs filling up . . . pulling in fresh air . . . and then pushing out the used air . . . so easily . . . comfortably giving your body the oxygen it needs . . . Now follow some of that oxygen up into your wonderful brain . . . that's where oxygen helps your brain take charge of everything you do and feel and think . . . Think of your brain as a super battery . . . sending electrical messages to all parts of your body . . . That really does happen inside of us . . . Tiny charges of electricity run our minds and our muscles . . . even as we rest here . . . getting quieter and quieter . . . We can imagine an electrical message from our brain to one toe . . . whichever toe you want to get the message to wiggle a bit inside your shoe . . . And the message goes back from that toe to the brain saying "Ahhhhhh . . . that felt good." . . . Now let's pretend to feel the energy in the circuit we just made . . . The energy we passed along with hand squeezes . . . Feel that energy starting from your brain . . . flowing down into your neck . . . out your shoulders . . . down into your arms and elbows and wrists and hands and fingertips . . . I wonder if you can really feel tiny tingles in your fingertips now? . . . and now . . . feel that charge of electricity flow down into your spine . . . and all around your ribs . . . and into your belly and hips . . . Now it's flowing through your legs . . . your knees . . . your ankles . . . your feet . . . your toes. Now as we are still and quiet . . . we feel that energy completing its circuit back to our brains . . . back to help our brains bring our thinking back to this room . . . to open our eyes and smile and stretch . . . and let ourselves feel fresh and ready to (do whatever comes next in the daily program at school).

Field Trips

Two simple field trips can extend young children's information about using electric power, at their level of understanding.

1. Ask a hardware store manager if he would show children electrical fittings such as switches, receptacles, and light sockets. He could point out where two wires must be connected to let the current pass through and where light bulbs must touch sockets to complete a circuit.
2. Walk to the utility pole nearest your school. Look for the insulators that keep the wires from touching the pole. If possible, follow the lines back to your building, checking to see if insulators are used where the wires might otherwise touch the building. If you should ever pass a telephone or electrical repair crew at work while walking with your children, stop to watch. Perhaps a repairman would be willing to show the children the equipment he wears on his belt, especially the rubber or plastic insulation on tool handles.

SECONDARY INTEGRATION

Keeping Concepts Alive

Whenever you notice the effects of static electricity, mention them to the children. "Did some of your hair seem to stand up when you were brushing it this morning? Mine did." The children will probably tell you why it happened. When you are ready to turn on a record player or a projector for use in school, ask what is needed to complete the circuit to make the electricity flow into the equipment. If your classroom has a computer, mention that many tiny switches, called *transistors*, control the flow of electricity to make it work. Talk about stormy weather when it occurs. "Were you awake during the storm last night? What did you see in the sky? Did you watch it a while? Did you remember how we made our tiny lightning sparks at school?"

Relating New Concepts to Existing Concepts

When the children are experimenting with static electricity charges, making unlike materials cling and like materials push away from each other, recall that the like ends of two magnets also pushed each other away and the unlike ends pulled together. Magnetism is somewhat like electricity in this respect (see page 199).

Relate concepts about the human body and nutrition to electricity concepts. Recall that our bodies get their energy from the food we eat. Some of this energy becomes the electricity that sends all the messages from the brain to our muscles and senses and back again. This important electrical communication system is our nervous system. Our brain also uses tiny electrical impulses to remember and think about things. Do the imagery exercise on page 273. Read *Thinking* by Kathie Smith and Victoria Crenson.

Family Involvement

Families can help their children look for things at home that use electricity. Suggest that they let their children watch the unloading of a clothes dryer. Do nylon synthetic garments cling to other things in the load? (They will unless a fabric softener was used.) Can the children hear crackling sounds when the clinging things are separated? If the area is not brightly lighted perhaps tiny sparks can be seen when the static electricity jumps.

Encourage children to discuss at home ways that families can save energy. (see: *50 Simple Things Kids Can Do to Save the Earth* in Resources).

EXPLORING AT HOME*

Sparks and Clings. Static electricity shows up more dramatically at home at night than it does during the school day. During dry weather, join your child in

*These suggestions may be duplicated and sent home to families.

scuffling your feet across the carpet in a fairly dark room, to build up a static charge. Touch a metal doorknob. Watch and listen for the spark that occurs. Touch one another and touch other objects to see whether static charges jump from you to them. Point out times when static might build as you or your child slide across a plastic seat getting into the car.

Turn down the lights near your dryer as you pull out a load of clothes that has *not* been treated with softener. (A load of jeans and polyester socks would be just fine for this.) Watch and listen for static sparks as you pull the clinging socks from the jeans. Watch for rice grains clinging inside a plastic bag as you pour it into a measuring cup. Let your child think about all those grains rubbing together, building up static charges inside the bag as they tumble out. Notice together how hair flies up toward the comb or brush when you are doing your hair on a dry day, how certain garments cling to your legs on those days.

Safety note: Clearly and often talk about safe use of electric appliances and extension cords. Emphasize the danger of using equipment plugged into powerful household current near *any* source of water or dampness.

RESOURCES

EARTHWORKS GROUP. *50 Simple Things Kids Can Do to Save the Earth.* Kansas City: Andrews & McMeel, 1990. Four of the conservation activities children can do at home focus on saving electricity. Recycled paperback.

MACAULAY, DAVID. *The Way Things Work.* Boston: Houghton Mifflin, 1988.

SMITH, KATHIE B., & CRENSON, VICTORIA. *Thinking*. Mahwah, NJ: Troll, 1988. This simple, excellent text on how the brain works includes a mention of the body's electrical "wiring." Paperback.

SPETGANG, TILLY, & WELLS, MALCOLM. *The Children's Solar Energy Book.* New York: Sterling, 1982.

APPENDIX 1

Resources for Music, Recordings, Poetry, and Creative Movement

Music

BOARDMAN, EUNICE, & LANDIS, BETH. *Exploring Music 1.* New York: Holt, Reinhart & Winston, 1969.
HAINES, JOAN, & GERBER, LINDA. *Leading Young Children to Music.* 3rd Edition. Columbus, OH: Merrill, 1988.
LANDECK, BEATRICE, & CROOK, ELIZABETH. *Wake Up and Sing.* New York: Edward Marks Music Corporation, William Morrow, 1969.
RAFFI. *Singable Songbook.* New York: Crown, 1987.
SEEGER, RUTH. *Animal Folk Songs for Children.* Garden City, NY: Doubleday, 1950.
SILBER, IRWIN. *Folksong Festival.* New York: Scholastic Book Services, 1974.

Recordings

BARLIN, ANNE LIEF. Dance-A-Story, *Balloons.* RCA Victor, LF104.
BERMAN, MARCIA. *Rabbits Dance: Marcia Berman Sings Malvina Reynolds.* B/B Records, 1985. 570 N. Arden Blvd., Los Angeles, CA 90004.
BURSTEIN, JOHN. *The Inside Story.* J. Burstein, Canal Street Station, Box 773, New York, NY 10013.
GLAZER, TOM. Singing *Now We Know (Songs to Learn By).* Columbia Records, CL670.
JENKINS, ELLA. Singing *Seasons For Singing.* Smithsonian audiocassette 45031.
ROGERS, FRED. *Won't You Be My Neighbor?* Family Communications, Inc. audiocassette MRV 8106C.
_____. *You Are Special.* audiocassette MRV 8103C.
SEEGER, PETE. Singing *Birds, Beasts, Bugs and Little Fishes.* Smithsonian audiocassette 45024.
ZAMFIR, GHEORGE. *Romance of the Pan Flute.* audiocassette #32150. Polygram Classics: 810 Seventh Ave., New York, N.Y.

Poetry

BROWN, MARGARET WISE. *Nibble Nibble.* New York: Young Scott Books, 1959.
DEREGNIERS, BEATRICE SCHENCK, MOORE, EVA, & WHITE, MARY M. *Poems Children Will Sit Still For.* New York: Scholastic Book Services, 1973.

FISHER, AILEEN. *When It Comes to Bugs.* New York: Harper & Row, 1986.
HOBAN, RUSSELL. *Egg Thoughts and Other Frances Songs.* New York: Harper & Row, 1974.
MILNE, A. A. *Now We Are Six.* New York: E. P. Dutton, 1961.
RYDER, JOANNE. *Inside Turtle's Shell, and Other Poems of the Field.* New York: Macmillan, 1985.
SILVERSTEIN, SHEL. *A Light In The Attic.* New York: Harper & Row, 1981.
SINGER, MARILYN. *Turtle In July.* New York: Macmillan, 1989.
STEVENSON, ROBERT LOUIS. *A Child's Garden of Verses.* New York: Airmont, 1969.
THOMPSON, JEAN MCKEE. *Poems To Grow On.* Boston: Beacon Press, 1957.

Creative Movement

CHERRY, CLARE. *Creative Movement for the Developing Child.* Palo Alto, CA: Fearon, 1968.
SHEEHY, EMMA. *Children Discover Music and Dance.* New York: Teachers College Press, 1968.
SINCLAIR, CAROLINE. *Movement of the Young Child Ages Two to Six.* Columbus, OH: Merrill, 1976. (out of stock; obtain copy from library.)

APPENDIX 2

Resources for Exploring at Home

Beane, Rona. *From Little Acorns to Mighty Oaks.* New York: Workman, 1989. Paperback.
Brown, Marc. *Your First Garden Book.* Boston: Little, Brown, 1981. Paperback.
Conaway, Judith. *More Science Secrets.* Mahwah, NJ: Troll, 1987. Paperback.
Gabrielson, Ira, & Herbert Zim. *Birds.* New York: Golden Books, 1987. Pocket guide.
Griffin, Sandra. *Earth Circles.* New York: Walker, 1989.
Lewis, James. *Learn While You Scrub: Science in the Tub.* New York: Simon & Schuster, 1989. Paperback.
McFarlane, Ruth. *Making Your Own Nature Museum.* New York: Franklin Watts, 1989.
Simon, Seymour. *Soap Bubble Magic.* New York: Lothrop, 1985.
Stokes, Donald and Lillian. *The Bird Feeder Book.* Boston: Little, Brown, 1987.
Van Cleave, Janice. *Chemistry for Every Kid: 101 Experiments That Really Work.* New York: John Wiley, 1989.
Waters, Gaby. *Science Tricks and Magic.* Tulsa: EDC Publishing, 1986. Paperback.
Wilkes, Angela. *My First Nature Book.* New York: Alfred A. Knopf, 1990.
Wyler, Rose. *Science Fun with Peanuts and Popcorn.* New York: Simon & Schuster, 1986. Paperback.
______. *Science Fun with Mud and Dirt.* New York: Simon & Schuster, 1986. Paperback.
______. *Science Fun with Toy Cars and Trucks.* New York: Simon & Schuster, 1988. Paperback.
Zim, Herbert, & Paul Shaffer. *Rocks and Minerals.* New York: Golden Books, 1989. Paperback pocket guidebook.

Free Resource

Coping With Children's Reactions to Earthquakes and Other Disasters.
Write to: Federal Emergency Management Agency, P.O. Box 70274, Washington, DC 20024. Ask for Bulletin FEMA #48.

Resources for Astronomy and Prehistoric Animals

Astronomy Books For Ages 4 Through 8

BRANLEY, FRANKLYN. *The Sky Is Full of Stars.* New York: Thomas Y. Crowell, 1981.
______. *Comets.* New York: Harper, 1984. Paperback.
______. *What the Moon Is Like.* New York: Harper, 1986. Paperback.
______. *The Moon Seems to Change.* New York: Harper, 1987. Paperback.
______. *The Planets in Our Solar System.* New York: Harper, 1987. Paperback. Lighthearted illustrations add appeal to this excellent series.
______. *Eclipse.* New York: Harper, 1988. Paperback.
______. *The Sun: Our Nearest Star.* New York: Harper, 1988. Paperback.
JACKSON, KIM. *Planets.* Mahwah, NJ: Troll, 1985. Paperback.
KAUFMAN, JOE. *Joe Kaufman's Big Book About Earth and Space.* New York: Golden Books, 1987.
SANTREY, LAURENCE. *Moon.* Mahwah, NJ: Troll, 1985. Paperback.
______. *Discovering the Stars.* Mahwah, NJ: Troll, 1982. Paperback.

Prehistoric Animal Books for Ages 4 Through 8

ALIKI, *Dinosaur Bones.* New York: Harper & Row, 1988. Chronicles how "dinosaur fever" began 150 years ago with the first reconstruction of a dinosaur skeleton from fossil bones. Paperback.
______. *Digging Up Dinosaurs.* New York: Harper & Row, 1988. Paperback.
BARTON, BYRON. *Dinosaurs, Dinosaurs.* New York: Thomas Y. Crowell, 1989. The minimal, large type text and simple concepts will appeal to the youngest learners and to beginning readers.
BERENSTAIN, MICHAEL. *The Biggest Dinosaurs.* New York: Golden Books, 1989. Paperback.
______. *The Horned Dinosaur: Triceratops.* New York: Golden Books, 1989. Paperback.
______. *The Spike-Tailed Dinosaur: Stegosaurus.* New York: Golden Books, 1989. Paperback.
______. *King of the Dinosaurs: Tyrannosaurus Rex.* New York: Golden Books, 1989. This series of well-written, inexpensive books offers interesting ideas about the likely behaviors of these prehistoric animals.
BRANLEY, FRANKLYN. *What Happened to the Dinosaurs?* New York: Thomas Y. Crowell, 1989. Older primary-grade children will be interested in pondering the theories about dinosaur extinction reported in this book.
COHEN, MIRIAM. *Lost in the Museum.* New York: Greenwillow, 1979. Dinosaur skeletons lure children on a natural history museum tour.
CRAIG, JANET. *Discovering Prehistoric Animals.* Mahwah, NJ: Troll, 1990. Information about varied prehistoric creatures who ruled land, sea, and air. For upper primary-grade children. Paperback.

ELTING, MARY. *Dinosaurs.* New York: Golden Books, 1987. Clear writing informs readers about making inferences from fossils to likely dinosaur behavior. It mentions the discoveries of a female paleontologist. Paperback.

______. *The Big Golden Book of Dinosaurs.* New York: Golden Books, 1988. Pleasant, flowing prose and well-organized information make this a good book to read aloud to children.

GIBBONS, GAIL. *Prehistoric Animals.* New York: Holiday House, 1988. Catalogs the wide variety of extinct animals. It traces the evolution of many animals living today, and identifies those that survive in their prehistoric form.

GRANGER, JUDITH. *Amazing World of Dinosaurs.* Mahwah, NJ: Troll, 1982. Four theories of dinosaur extinction are reported. Paperback.

INGOGLIA, GINA. *Those Mysterious Dinosaurs.* New York: Golden Books, 1989. For young learners. Paperback.

MILTON, JOYCE. *Dinosaur Days.* New York: Random House, 1985. This book clarifies the separate existence of prehistoric animals and human life. A simple text for beginning readers. Paperback.

MORT, BERNARD. *The Littlest Dinosaur.* New York: Harcourt, Brace, Jovanovich, 1989. Not all dinosaurs were huge. The sizes of small ones are compared to familiar objects like the sofa, the tub, or the teddy bear.

PETERSON, DAVE. *Apatosaurus.* Chicago: Children's Press, 1989. A mix-up gave this dinosaur the incorrect name of Brontosaurus.

SIMON, SEYMOUR. *New Questions and Answers About Dinosaurs.* New York: Morrow, 1990. The text is framed as answers to questions that children often ask about dinosaurs.

APPENDIX 4

Salvage Sources for Science Materials

Resourceful teachers know how to compensate for tight equipment budgets. They scout the community for discarded materials that can be put to use in their classrooms. Most of the materials and some of the equipment called for in this book were acquired in this way. As our society shifts from a throw-away orientation to one of more conservative use of materials, some of these items may become scarce. Depend upon personal ingenuity to devise even better substitutes for the obsolete materials. The following is a list of the current suppliers of the materials on science shelves.

Clinic medical laboratory:

- discarded test-kit bottles with dropper tops (ask if sterilizing is advisable before using in school)
- plastic prescription vials with caps
- screw-top plastic bottles
- discarded syringes (needles removed)

Dress shop alterations department or drapery shop:

- empty thread spools

Electric motor repair shop:

- old magnets from motors

Grocery store:

- plastic berry baskets (available during fresh berry season)

Photography studio:

- color film cannisters, film spools, 35mm film cans, small plastic chemical bottles (wash carefully before reusing)

Retail shops of all sorts:

- foam packing chips, molded foam packing sheets, plastic bubble packing papers

Shoe store:

- shoe boxes (our science supplies are stored in 26 shoe boxes)

Telephone installation crews:

- short pieces of telephone cable containing copper wires insulated in nice colors

Many parents are willing to support science activities by saving materials or lending equipment. A list of appreciated materials might include these:

- Aluminum foil pans
- Bottle caps
- Broken hand-wound clock
- Broomsticks
- Coffee cans with plastic lids
- Corn popper with clear plastic dome
- Cottage cheese cartons
- Disposable plastic tumblers
- Fabric scraps
- Frozen juice cans
- Garden hose lengths (clear)
- Ice cube trays of molded plastic
- Lids from cocoa tins
- Margarine tubs
- Meat trays (foam or clear plastic)
- Mesh onion or potato bags
- Milkshake container lids (plastic)
- Mirrors (pocket)
- Newspapers
- Plastic prescription vials, screw-top bottles
- Muffin pans
- Paper bags
- Pulleys
- Small, rigid suitcase
- 2-liter plastic soda bottles
- Washers (metal)
- Seed catalogs
- Vitamin drop bottle (plastic)
- Carry-out covered food containers

Families may be willing to lend:

- Acrobat balance toy
- Cookie press (screw-type)
- Bicycle tire pump
- Camper's step pump (bellows type)
- Kitchen scale
- Postage scale, or calorie-count scale
- Old fashioned piano stool
- $1\frac{1}{2}$ volt dry cell (#6). (If not available, consider purchasing "D" flashlight battery holders from the SEE catalog to take the place of the more expensive dry cell. Purchase "D" batteries locally.)

Reasonably priced, sturdy science equipment for classroom use is sold by mail from:

(*Science and Things* catalog.)
Selective Educational Equipment (SEE), Inc.
3 Bridge St.
Newton, MA 02195

(Discount supply catalog.)
American Science Center
5430 W. Layton Ave.
Milwaukee, WI 53220

HearthSong, A Catalog for Families
P. O. Box B
Sebastopol, CA 95473

Seventh Generation, Products for a Healthy Planet
Colchester, VT 05446–1672

The Nature Company Catalog
P. O. Box 2310
Berkeley, CA 94702

Lawrence Hall of Science
University of California
Berkeley, CA 94270

Index